I0502853

Status and Understanding of Groundwater Quality in the Madera-Chowchilla Study Unit, 2008: California GAMA Priority Basin Project

By Jennifer L. Shelton, Miranda S. Fram, Kenneth Belitz, and Bryant C. Jurgens

A product of the California Groundwater Ambient Monitoring and Assessment (GAMA) Program

Prepared in cooperation with the California State Water Resources Control Board

Scientific Investigations Report 2012–5094

U.S. Department of the Interior
U.S. Geological Survey

U.S. Department of the Interior
KEN SALAZAR, Secretary

U.S. Geological Survey
Marcia K. McNutt, Director

U.S. Geological Survey, Reston, Virginia: 2013

For more information on the USGS—the Federal source for science about the Earth, its natural and living resources, natural hazards, and the environment, visit http://www.usgs.gov or call 1–888–ASK–USGS.

For an overview of USGS information products, including maps, imagery, and publications, visit http://www.usgs.gov/pubprod

To order this and other USGS information products, visit http://store.usgs.gov

Suggested citation:
Shelton, J.L., Fram, M.S., Belitz, Kenneth, and Jurgens, B.C., 2013, Status and understanding of groundwater quality in the Madera-Chowchilla Study Unit, 2008—California GAMA Priority Basin Project: U.S. Geological Survey Scientific Investigations Report 2012–5094, 86 p.

Contents

Contents—Continued

Figures

Figures—Continued

Figures—Continued

Tables

Tables—Continued

Conversion Factors, Datums, and Abbreviations and Acronyms

Conversion Factors

Inch/Pound to SI

Multiply	By	To obtain
	Length	
inch (in.)	2.54	centimeter (cm)
inch (in.)	25.4	millimeter (mm)
foot (ft)	0.3048	meter (m)
mile (mi)	1.609	kilometer (km)
	Area	
square foot (ft^2)	929.0	square centimeter (cm^2)
square foot (ft^2)	0.09290	square meter (m^2)
square mile (mi^2)	259.0	hectare (ha)
square mile (mi^2)	2.590	square kilometer (km^2)
	Radioactivity	
picocurie per liter (pCi/L)	0.037	becquerel per liter (Bq/L)

Temperature in degrees Celsius (°C) may be converted to degrees Fahrenheit (°F) as follows:

$$°F=(1.8 \times °C)+32$$

Temperature in degrees Fahrenheit (°F) may be converted to degrees Celsius (°C) as follows:

$$°C=(°F-32)/1.8$$

Datums

Vertical coordinate information is referenced to the North American Vertical Datum of 1988 (NAVD 88).

Horizontal coordinate information is referenced to the North American Datum of 1983 (NAD 83).

Elevation, as used in this report, refers to distance above the vertical datum.

Specific conductance is given in microsiemens per centimeter at 25 degrees Celsius (μS/cm at 25 °C).

Concentrations of chemical constituents in water are given either in milligrams per liter (mg/L) or micrograms per liter (μg/L).

Conversion Factors, Datums, and Abbreviations and Acronyms—Continued

Abbreviations and Acronyms

AB	Assembly Bill (through the California State Assembly)
AL-US	U.S. Environmental Protection Agency action level
BQ	benchmark quotient
Fract-CaMg	calcium plus magnesium in milliequivalents divided by the sum of calcium, magnesium, sodium, and potassium in milliequivalents
GAMA	Groundwater Ambient Monitoring and Assessment Program
HAL-US	U.S. Environmental Protection Agency lifetime health advisory level
HBSL	health-based screening level
MADCHOW	prefix for Madera-Chowchilla study unit grid well
MADCHOWFP	prefix for Madera-Chowchilla study unit understanding well
MCL-US	U.S. Environmental Protection Agency maximum contaminant level
NL-CA	California Department of Public Health notification level
pmc	percent modern carbon
redox	oxidation-reduction
RC	relative-concentration
RSD5-US	U.S. Environmental Protection Agency risk-specific dose at risk factor of 10^{-5}
SMCL	secondary maximum contaminant level
SMCL-CA	California Department of Public Health secondary maximum contaminant level
SMCL-US	U.S. Environmental Protection Agency secondary maximum contaminant level
TEAP	terminal electron acceptor processes
U.S.	United States

Organizations

CDPH	California Department of Public Health (was California Department of Health Services prior to July 1, 2007)
CDWR	California Department of Water Resources
LLNL	Lawrence Livermore National Laboratory
NAWQA	National Water-Quality Assessment Program (USGS)
SWRCB	State Water Resources Control Board (California)
USEPA	U.S. Environmental Protection Agency
USGS	U.S. Geological Survey

Conversion Factors, Datums, and Abbreviations and Acronyms—Continued

Selected Chemical Names

^{14}C	carbon-14
DO	dissolved oxygen
DBCP	1,2-dibromo-3-chloropropane
EDB	1,2-dibromoethane
D-D	dichloropropane-dichloropropene mixture
He	helium
^{3}He	helium-3
^{4}He	helium-4
NDMA	*N*-nitrosodimethylamine
PCE	tetrachloroethene
1,2,3-TCP	1,2,3-trichloropropane
TDS	total dissolved solids
THM	trihalomethane
^{3}H	tritium
VOC	volatile organic compound

Units of Measure

cm^3 STP g^{-1}	centimeters at standard temperature and pressure per gram
L	liter
mg/L	milligrams per liter (parts per million)
μg/L	micrograms per liter (parts per billion)
μS/cm	microsiemens per centimeter
per mil	parts per thousand
ppb	parts per billion
TU	tritium unit
>	greater than
≥	greater than or equal to
<	less than
≤	less than or equal to
—	not detected
%	percent

Status and Understanding of Groundwater Quality in the Madera-Chowchilla Study Unit, 2008: California GAMA Priority Basin Project

By Jennifer L. Shelton, Miranda S. Fram, Kenneth Belitz, and Bryant C. Jurgens

Abstract

Groundwater quality in the approximately 860-square-mile Madera and Chowchilla Subbasins (Madera-Chowchilla study unit) of the San Joaquin Valley Basin was investigated as part of the Priority Basin Project of the Groundwater Ambient Monitoring and Assessment (GAMA) Program. The study unit is located in California's Central Valley region in parts of Madera, Merced, and Fresno Counties. The GAMA Priority Basin Project is being conducted by the California State Water Resources Control Board in collaboration with the U.S. Geological Survey (USGS) and the Lawrence Livermore National Laboratory. The Project was designed to provide statistically robust assessments of untreated groundwater quality within the primary aquifer systems in California. The primary aquifer system within each study unit is defined by the depth of the perforated or open intervals of the wells listed in the California Department of Public Health (CDPH) database of wells used for municipal and community drinking-water supply. The quality of groundwater in shallower or deeper water-bearing zones may differ from that in the primary aquifer system; shallower groundwater may be more vulnerable to contamination from the surface.

The assessments for the Madera-Chowchilla study unit were based on water-quality and ancillary data collected by the USGS from 35 wells during April–May 2008 and water-quality data reported in the CDPH database. Two types of assessments were made: (1) *status*, assessment of the current quality of the groundwater resource, and (2) *understanding*, identification of natural factors and human activities affecting groundwater quality. The primary aquifer system is represented by the grid wells, of which 90 percent (%) had depths that ranged from about 200 to 800 feet (ft) below land surface and had depths to the top of perforations that ranged from about 140 to 400 ft below land surface.

Relative-concentrations (sample concentrations divided by benchmark concentrations) were used for evaluating groundwater quality for those constituents that have Federal or California regulatory or non-regulatory benchmarks for drinking-water quality. A relative-concentration (RC) greater than 1.0 indicates a concentration above a benchmark. RCs for organic constituents (volatile organic compounds and pesticides) and special-interest constituents (perchlorate) were classified as "high" (RC is greater than 1.0), "moderate" (RC is less than or equal to 1.0 and greater than 0.1), or "low" (RC is less than or equal to 0.1). For inorganic constituents (major and minor ions, trace elements, nutrients, and radioactive constituents), the boundary between low and moderate RCs was set at 0.5. The assessments characterize untreated groundwater quality, not the quality of treated drinking water delivered to consumers by water purveyors; drinking-water benchmarks, and thus relative-concentrations, are used to provide context for the concentrations of constituents measured in groundwater.

Aquifer-scale proportion was used in the status assessment as the primary metric for evaluating regional-scale groundwater quality. High aquifer-scale proportion is defined as the percentage of the area of the primary aquifer system with RCs greater than 1.0 for a particular constituent or class of constituents; moderate and low aquifer-scale proportions are defined as the percentages of the area of the primary aquifer system with moderate and low RCs, respectively. Percentages are based on an areal, rather than a volumetric basis. Two statistical approaches—grid-based, which used one value per grid cell, and spatially weighted, which used multiple values per grid cell—were used to calculate aquifer-scale proportions for individual constituents and classes of constituents. The spatially weighted estimates of high aquifer-scale proportions were within the 90% confidence intervals of the grid-based estimates for all constituents except iron.

The *status assessment* showed that inorganic constituents had greater high and moderate aquifer-scale proportions in the Madera-Chowchilla study unit than did organic constituents. RCs for inorganic constituents with health-based benchmarks were high in 37% of the primary aquifer system, moderate in 30%, and low in 33%. The inorganic constituents contributing most to the high aquifer-scale proportion were arsenic (13%), uranium (17%), gross alpha particle activity (20%), nitrate (6.7%), and vanadium (3.3%). RCs for inorganic constituents with non-health-based benchmarks were high in 6.7% of the primary aquifer system, and the constituent contributing most to the high aquifer-scale proportion was total dissolved solids (TDS). RCs for organic constituents with health-based benchmarks were high in 10% of the primary aquifer system, moderate in 3.3%, and low in 40%; organic constituents were not detected in 47% of the primary aquifer system. The fumigant 1,2-dibromo-3-chloropropane (DBCP) was the only organic constituent detected at high RCs. Seven organic constituents were detected in 10% or more of the primary aquifer system: DBCP; the fumigant additive 1,2,3-trichloropropane; the herbicides simazine, atrazine, and diuron; the trihalomethane chloroform; and the solvent tetrachloroethene (PCE). RCs for the special-interest constituent perchlorate were moderate in 20% of the primary aquifer system.

The second component of this study, the *understanding assessment*, identified the natural and human factors that may affect groundwater quality by evaluating statistical correlations between water-quality constituents and potential explanatory factors, such as land use, position relative to important geologic features, groundwater age, well depth, and geochemical conditions in the aquifer. Results of the statistical evaluations were used to explain the distribution of constituents in the study unit. Depth to the top of perforations in the well and groundwater age were the most important explanatory factors for many constituents. High and moderate RCs of nitrate, uranium, and TDS and the presence of herbicides, trihalomethanes, and solvents were all associated with depths to the top of perforations less than 235 ft and modern- and mixed-age groundwater. Positive correlations between uranium, bicarbonate, TDS, and the proportion of calcium and magnesium in the total cations suggest that downward movement of recharge from irrigation water contributed to the elevated concentrations of these constituents in the primary aquifer system. High and moderate RCs of arsenic were associated with depths to the top of perforations greater than 235 ft, mixed- and pre-modern-age groundwater, and location in sediments from the Chowchilla River alluvial fan, suggesting that increased residence time and appropriate aquifer materials were needed for arsenic to accumulate in the groundwater. High and moderate RCs of fumigants were associated with depths to the top of perforations of less than

235 ft and location south of the city of Madera; low RCs of fumigants were detected in wells dispersed across the study unit with a range of depths to top of perforations.

Introduction

Groundwater composes nearly half of the water used for public and domestic drinking-water supply in California (Kenny and others, 2009). To assess the quality of ambient groundwater in aquifers used for drinking-water supply and to establish a baseline groundwater quality monitoring program, the California State Water Resources Control Board (SWRCB), in collaboration with the U.S. Geological Survey (USGS) and Lawrence Livermore National Laboratory (LLNL), implemented the Groundwater Ambient Monitoring and Assessment (GAMA) Program (California State Water Resources Control Board, 2010, website at http://www.swrcb.ca.gov/gama/). The statewide GAMA Program currently consists of four projects: (1) the GAMA Priority Basin Project, conducted by the USGS (website at http://ca.water.usgs.gov/gama/); (2) the GAMA Domestic Well Project, conducted by the SWRCB; (3) the GAMA Special Studies, conducted by LLNL; and (4) the GeoTracker GAMA online database, conducted by the SWRCB. On a statewide basis, the GAMA Priority Basin Project focused primarily on the deeper portion of the groundwater resource, and the SWRCB Domestic Well Project generally focused on the shallower aquifer systems.

The SWRCB initiated the GAMA Program in 2000 in response to a legislative mandate (State of California, 1999, 2001a, Supplemental Report of the 1999 Budget Act 1999–00 Fiscal Year). The GAMA Priority Basin Project was initiated in response to the Groundwater Quality Monitoring Act of 2001 to assess and monitor the quality of groundwater in California (State of California, 2001b, Sections 10780–10782.3 of the California Water Code, Assembly Bill 599). The GAMA Priority Basin Project is a comprehensive assessment of statewide groundwater quality designed to help better understand and identify risks to groundwater resources and to increase the availability of information about groundwater quality to the public. For the GAMA Priority Basin Project, the USGS, in collaboration with the SWRCB, developed a monitoring plan to assess groundwater basins through direct sampling of groundwater and other statistically reliable sampling approaches (Belitz and others, 2003; California State Water Resources Control Board, 2003). Additional partners in the GAMA Priority Basin Assessment include the California Department of Public Health (CDPH), California Department of Pesticide Regulation (CDPR), California Department of Water Resources (CDWR), and local water agencies and well owners (Kulongoski and Belitz, 2004).

The range of hydrologic, geologic, and climatic conditions that exists in California must be considered in a statewide assessment of groundwater quality. Belitz and others (2003) partitioned the State into 10 hydrogeologic provinces, each with distinctive hydrologic, geologic, and climatic characteristics (fig. 1). All of these hydrogeologic provinces include groundwater basins and subbasins designated by the CDWR (California Department of Water Resources, 2003). Groundwater basins generally consist of relatively permeable, unconsolidated deposits of alluvial or volcanic origin. Eighty percent of California's approximately 16,000 active and standby drinking-water wells listed in the statewide database maintained by the CDPH (hereinafter referred to as CDPH wells) are located in CDWR-designated groundwater basins within these hydrogeologic provinces. Groundwater basins and subbasins were prioritized for sampling on the basis of the number of CDPH wells in the basin, with secondary consideration given to municipal groundwater use, agricultural pumping, the number of historically leaking underground fuel tanks, and registered pesticide applications (Belitz and others, 2003). Of the 472 basins and subbasins designated by the CDWR, 116 priority basins, representing approximately 95 percent (%) of the CDPH wells located in basins, were selected for the project. Some areas outside of the defined groundwater basins were also included to represent the approximately 20% of CDPH wells not located in groundwater basins. The 116 priority basins and additional areas outside of the defined groundwater basins were grouped into 35 study units.

The goal of the GAMA Priority Basin Project is to produce three types of water-quality assessments for each study unit—(1) *Status:* assessment of the current quality of the groundwater resource, (2) *Understanding:* identification of the natural and human factors affecting groundwater quality, and (3) *Trends:* detection of changes in groundwater quality (Kulongoski and Belitz, 2004). The assessments are intended to characterize the quality of groundwater within the primary aquifer system of the study unit, not the treated drinking water delivered to consumers by water purveyors. The primary aquifer systems for the study units are defined by the depths of the perforated or open intervals of the wells listed in the CDPH databases for the study units. The CDPH database lists wells used for municipal and community drinking-water supplies and includes wells from systems classified as non-transient (such as cities, towns, and mobile-home parks) and transient (such as schools, campgrounds, and restaurants). Groundwater quality in the primary aquifer system may differ from water in shallower or deeper parts of the aquifer systems. In particular, shallower groundwater may be more vulnerable to contamination from the land surface.

Purpose and Scope

The purposes of this report are to provide (1) a *study unit description*: briefly describe the hydrogeologic setting of the Madera-Chowchilla study unit, (2) a *status assessment*: assessment of the status of the current (2008) quality of groundwater in the primary aquifer system in the Madera-Chowchilla study unit, and (3) an *understanding assessment*: identification of the natural and human factors affecting groundwater quality, and an explanation of the relations between water quality and selected explanatory factors. In the Madera-Chowchilla study unit, the primary aquifer system corresponds to a depth interval of approximately 140 to 800 feet (ft) below land surface. Water-quality data for samples collected by the USGS for the GAMA Program in the Madera-Chowchilla study unit and details of sample collection, analysis, and quality-assurance procedures are described by Shelton and others (2009). Untreated groundwater samples were collected between April and May 2008. Utilizing those same data, this report describes methods used in designing the sampling network, identifying CDPH data for use in the status assessment, estimating aquifer-scale proportions for constituents, analyzing ancillary datasets, and assessing the status and understanding of groundwater quality by statistical and graphical approaches.

The *status assessment* uses two methods for calculating the areal proportion of the primary aquifer system with groundwater of defined quality (aquifer-scale proportion). Both methods are based on equal-area grid cells covering the study unit: one uses one well to represent each cell, and the other uses multiple wells to represent each cell. The first method is based on water-quality data from 30 wells selected by the USGS for spatial coverage of one well per grid cell across the study unit (grid wells). Samples were collected in April and May 2008 by the USGS for the GAMA Program for analysis of anthropogenic organic constituents, naturally occurring inorganic constituents, and geochemical and age-dating tracers (Shelton and others, 2009). The resulting set of water-quality data from the 30 grid wells was considered to be representative of the primary aquifer system in the Madera-Chowchilla study unit. The second method uses the water-quality data from the grid wells, water-quality data from 5 additional wells sampled by the USGS for the GAMA Program (understanding wells), and data reported for wells in the CDPH database during the most recent 3 years available at the time of the USGS sampling for the GAMA Program. GAMA *status assessments* are designed to provide a statistically robust characterization of groundwater quality in the primary aquifer systems at the study-unit scale (Belitz and others, 2003). The statistically robust design also allows study units to be compared and results to be synthesized at regional and statewide scales.

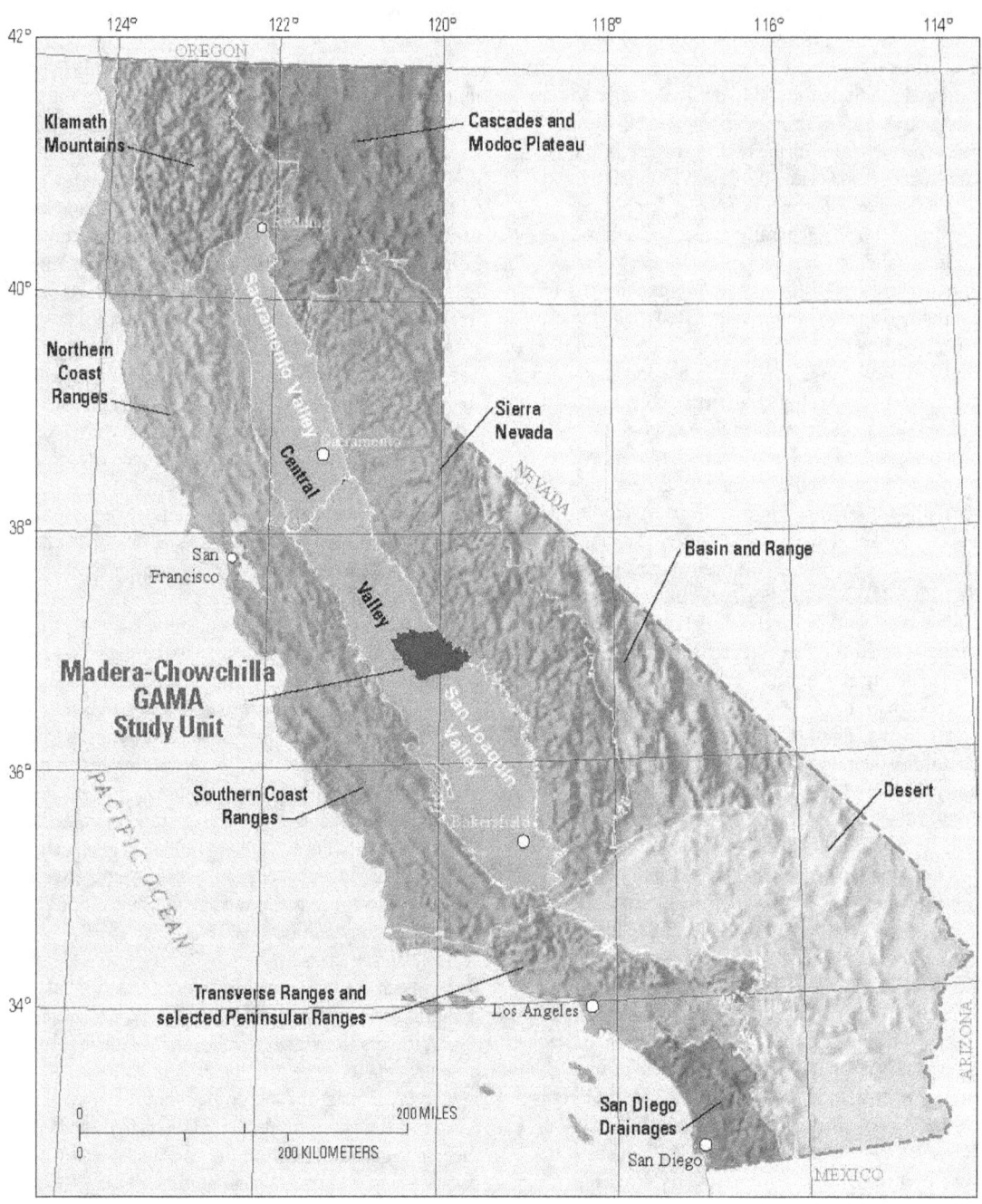

Figure 1.   Locations of the Madera-Chowchilla California Groundwater Ambient Monitoring and Assessment (GAMA) study unit and the California hydrogeologic provinces.

To provide context, the water-quality data discussed in this report are compared to California and Federal drinking-water regulatory and non-regulatory benchmarks for treated drinking water. Groundwater quality is defined in terms of relative-concentrations (the ratio of the concentration of a constituent in groundwater to the concentration of the benchmark for that constituent). The assessments in this report characterize the quality of untreated groundwater resources in the primary aquifer system in the study unit, not the treated drinking water delivered to consumers by water purveyors. This study does not attempt to evaluate the quality of water delivered to consumers; after withdrawal from the ground, water typically is treated, disinfected, and (or) blended with other waters to maintain acceptable water quality. Regulatory benchmarks apply to treated water that is delivered to the consumer, not to untreated groundwater.

The *understanding assessment* is based on water-quality data from the 30 grid wells and the 5 understanding wells sampled by the USGS for the GAMA Program. The potential explanatory factors affecting water quality in the primary aquifer system evaluated are land use, well depth, depth to the top of perforation, depth relative to the position of the Corcoran Clay, lateral position within the groundwater flow system, indicators of groundwater age, and geochemical conditions. Connections between potential explanatory factors and water quality are evaluated using statistical tests for correlations and by analysis of graphical relations.

Hydrogeologic Setting

The southern two-thirds of the Central Valley Hydrogeologic Province consists of the San Joaquin Valley (fig. 1). The Madera-Chowchilla study unit is composed of two subbasins of the CDWR San Joaquin Valley groundwater basin: Madera and Chowchilla (California Department of Water Resources, 2003). The study unit is bounded partially on the north by the Chowchilla River, approximately on the west and south by the San Joaquin River, and on the east by foothills of the Sierra Nevada (fig. 2).

Similar to most areas in the San Joaquin Valley of California, the Madera-Chowchilla study unit has a Mediterranean climate, with hot, dry summers and cool, moist winters. Average annual rainfall across the study unit ranges from 11 inches (in.) over most of the study unit to 15 in. in the eastern portions of the study unit along the foothills of the Sierra Nevada (Western Regional Climate Center, 2009).

The main surface-water features in the Madera-Chowchilla study unit are the San Joaquin, Fresno, and Chowchilla Rivers, and the Friant–Kern, Madera, and Chowchilla canals (fig. 2). There are also approximately 150 miles of irrigation pipelines, 300 miles of open-flow canal systems each supplying water to one or more farms, and numerous irrigation ditches (California Department of Water Resources, 1966; Madera Irrigation District, 2004). The San Joaquin, Fresno, and Chowchilla Rivers are dammed and have reservoirs in the Sierra Nevada foothills east of the study unit. The canals and pipelines deliver surface water from the major rivers and reservoirs to agricultural areas in the study unit.

The Madera-Chowchilla study unit is located on the eastern side of the San Joaquin Valley (fig. 1). The San Joaquin Valley is a structural trough 200 miles long and up to 70 miles wide, and is filled with up to 32,000 ft of marine and continental sediments that range in age from Jurassic to Holocene. The freshwater aquifer systems primarily are contained in the Late Tertiary and Quaternary continental deposits on the top of the pile (Page, 1986). These deposits increase in thickness from north to south and are up to 3,000 ft thick in the Madera-Chowchilla study unit. The continental deposits consist of alluvial fan and fluvial deposits with some interbedded lacustrine deposits. Most of the sediments were derived from the Sierra Nevada to the east, with lesser amounts of sediment derived from the Coast Ranges to the west. Three physiographic regions are defined in the Valley: the eastern alluvial fans, the western alluvial fans, and the basin in the center. Sediments consist of gravels, sands, silts, and clays, and generally are coarser at the proximal sides of the fans, closest to the Sierra Nevada and Coast Ranges, and become finer towards the center the basin (Gronberg and others, 1998). The most extensive lacustrine deposit, the Corcoran Clay Member of the Tulare Formation, underlies large parts of the western alluvial fans and basin, and the distal end of parts of the eastern alluvial fans at depths dipping from 50 ft on the eastern edge of the Clay to 300 ft along the margin of the Coast Ranges. The Corcoran Clay divides the San Joaquin Valley freshwater aquifer systems into an unconfined to semi-confined upper system and a largely confined lower system. The Madera-Chowchilla study unit includes eastern alluvial fan and basin areas (fig. 3).

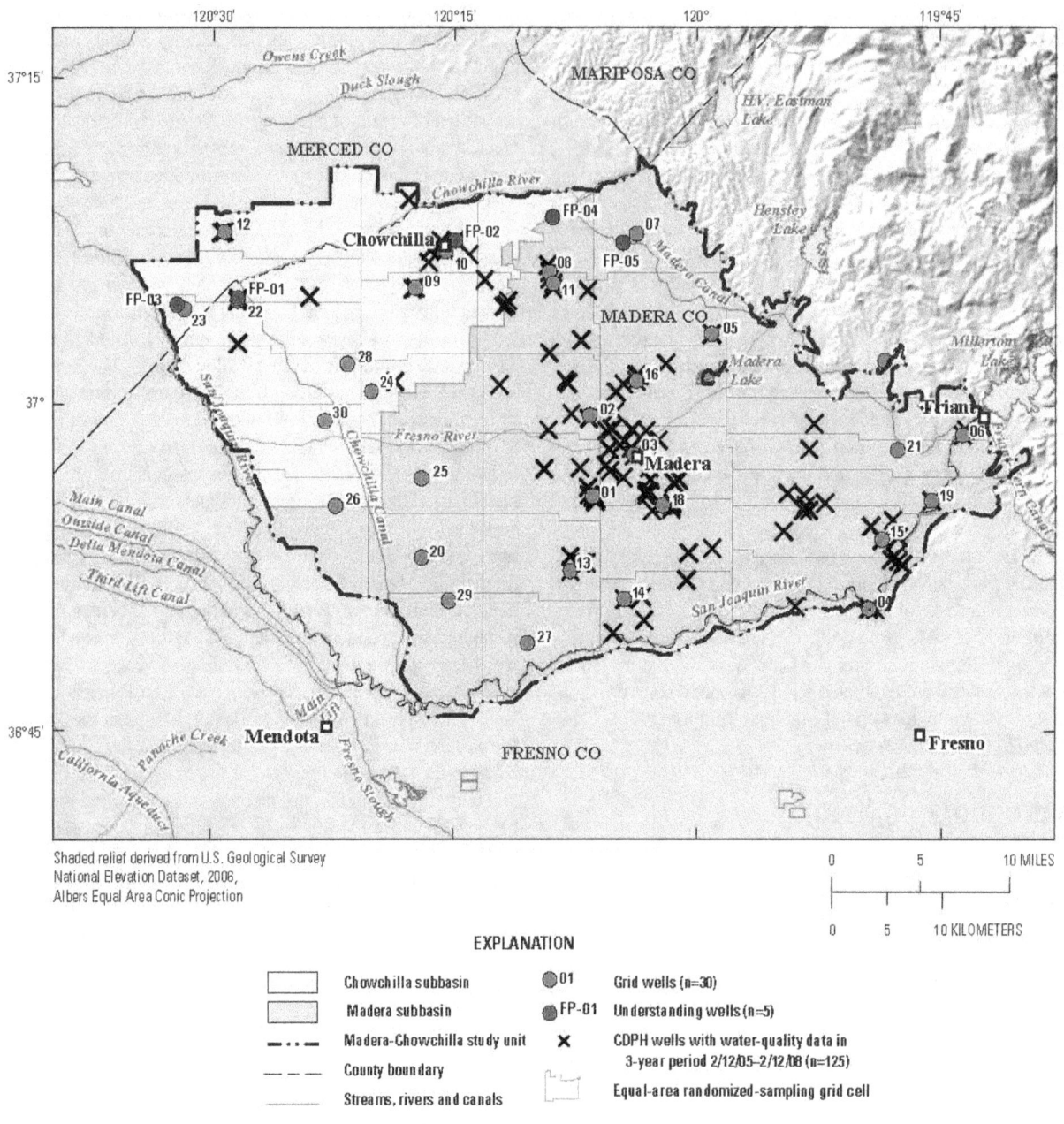

Figure 2. Geographic features and locations of grid cells, grid and understanding wells sampled during April–May 2008, and wells with data in the California Department of Public Health database during February 2005–February 2008, Madera-Chowchilla study unit, California GAMA Priority Basin Project.

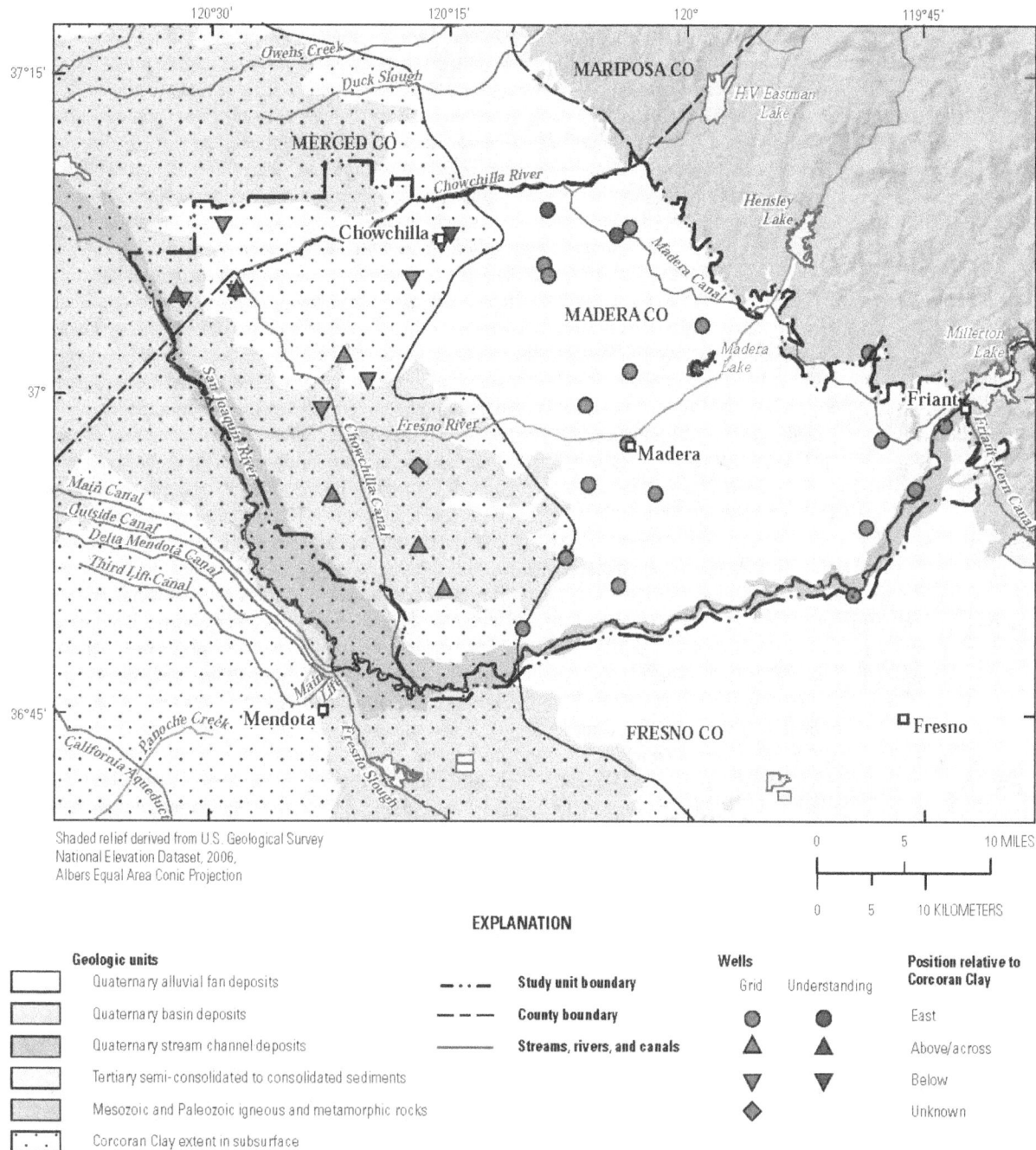

Figure 3. Madera-Chowchilla study unit, California GAMA Priority Basin Project.

Sediments in the eastern San Joaquin Valley primarily were deposited by large rivers draining glaciated areas of the Sierra Nevada (Mokelumne, Stanislaus, Tuolumne, Merced, San Joaquin, Kings, and Kaweah Rivers), and primarily are composed of material derived from granitic rocks of the Sierra Nevada batholith. The Madera-Chowchilla area is an exception because the Fresno and Chowchilla Rivers primarily have lower elevation watersheds that do not include glaciated areas of the Sierra Nevada (Weissmann and others, 2005). This difference affects the composition and depositional patterns of sediment in the study unit. The watershed of the Chowchilla River consists of a mixture of Mesozoic and Paleozoic metasedimentary and metavolcanic rocks, mafic intrusive rocks, and granitic rocks of the Sierra Nevada Batholith (Jennings, 1977; Saucedo and others, 2000; Weissmann and others, 2005), which results in mineralogically different sediment than sediment from granitic source rocks alone. These mineralogical differences are reflected in the soils, which have been extensively mapped in the Eastern San Joaquin Valley (Marchand and Allwardt, 1981). Soils derived from these mixed sources have a higher proportion of weatherable dark minerals (iron-magnesium silicate minerals and oxide and sulfide accessory minerals) and a lower proportion of quartz, compared to soils formed on sediments derived from granitic sources alone (Huntington, 1971). These differences are important because they may affect groundwater chemistry. The differences in sediment source also affect the physical structure of the sedimentary deposits. The glaciated areas of the Sierra Nevada yield much larger volumes of sediment than do the unglaciated areas; thus, the depositional sequences in the alluvial fans of rivers with headwaters in the glaciated areas are much thicker than those in the alluvial fans of rivers, such as the Chowchilla and Fresno Rivers, whose headwaters are in the unglaciated foothills (Weissmann and others, 2005).

The conceptual model of groundwater flow within the Madera-Chowchilla study unit (figs. 4A,B) is based on previous investigations in the Eastern San Joaquin Valley by Burow and others (2004) and Phillips and others (2007). Regional lateral flow of groundwater on the eastern side of the San Joaquin Valley and within the study unit is towards the southwest along the dip of the water-bearing units, and groundwater flows generally towards the axial trough (fig. 4A). Irrigation return flows are the major source of groundwater recharge, and groundwater pumping is the major source of discharge (Mitten and others, 1970; California Department of Water Resources, 2004a,b; Faunt, 2009). Groundwater on a lateral flow path may be repeatedly extracted by pumping wells and reapplied at the surface multiple times before reaching the valley trough (Phillips and others, 2007). This recharge and discharge pattern results in a substantial component of downward vertical flow (fig. 4B; Burow and others, 2004; Phillips and others, 2007; Faunt, 2009). These vertical flow components enhance vertical movement of water from recharge areas to the perforated intervals of withdrawal wells within shallow to intermediate depths in the system. These processes may occur in both agricultural and urban land-use areas. Groundwater age is vertically stratified, with water less than 50 years old in the upper parts of the aquifer system and water that may be tens of thousands of years old at depth (Burow and others, 2008). In the western part of the study unit, the Corcoran Clay may restrict the interaction between underlying confined and overlying unconfined groundwater; however, well-bores open to the aquifer above and below the Clay permit water exchange across the units (Williamson and others, 1989). At the western end of the flow system, there is upward movement of groundwater towards the San Joaquin River. Groundwater flow was artesian in the western portion of the study unit (Mendenhall, 1908; Mendenhall and others, 1916), but with increasing groundwater development, water levels have generally exhibited a long-term declining trend (Todd Engineers, 2002; California Department of Water Resources, 2004a,b).

Land use in the study unit is approximately 69% agricultural, 28% natural, and 3% urban on the basis of classification of USGS National Land Cover Data (Nakagaki and others, 2007) (fig. 5A). The agricultural land-use areas are mostly vineyards and orchards. Agricultural land use is distributed over most of the study unit (fig. 6), and natural land use occurs primarily along the eastern margin and in the central part of the western portion of the study unit. Most of the natural areas in the western portion of the study unit are grasslands and wetlands used as wildlife preserves. Most of the study unit is in Madera County (fig. 2), one of the fastest growing counties in the State (U.S. Census Bureau, 2010). The largest urban areas within the Madera-Chowchilla study unit are the cities of Madera and Chowchilla. The population of the city of Madera in July 2009 was 56,692, an increase of 31% since 2000 (U.S. Census Bureau, 2010). The population of the city of Chowchilla was 19,254 in July 2009, an increase of 73% since 2000. The city of Fresno is just to the south of the study unit.

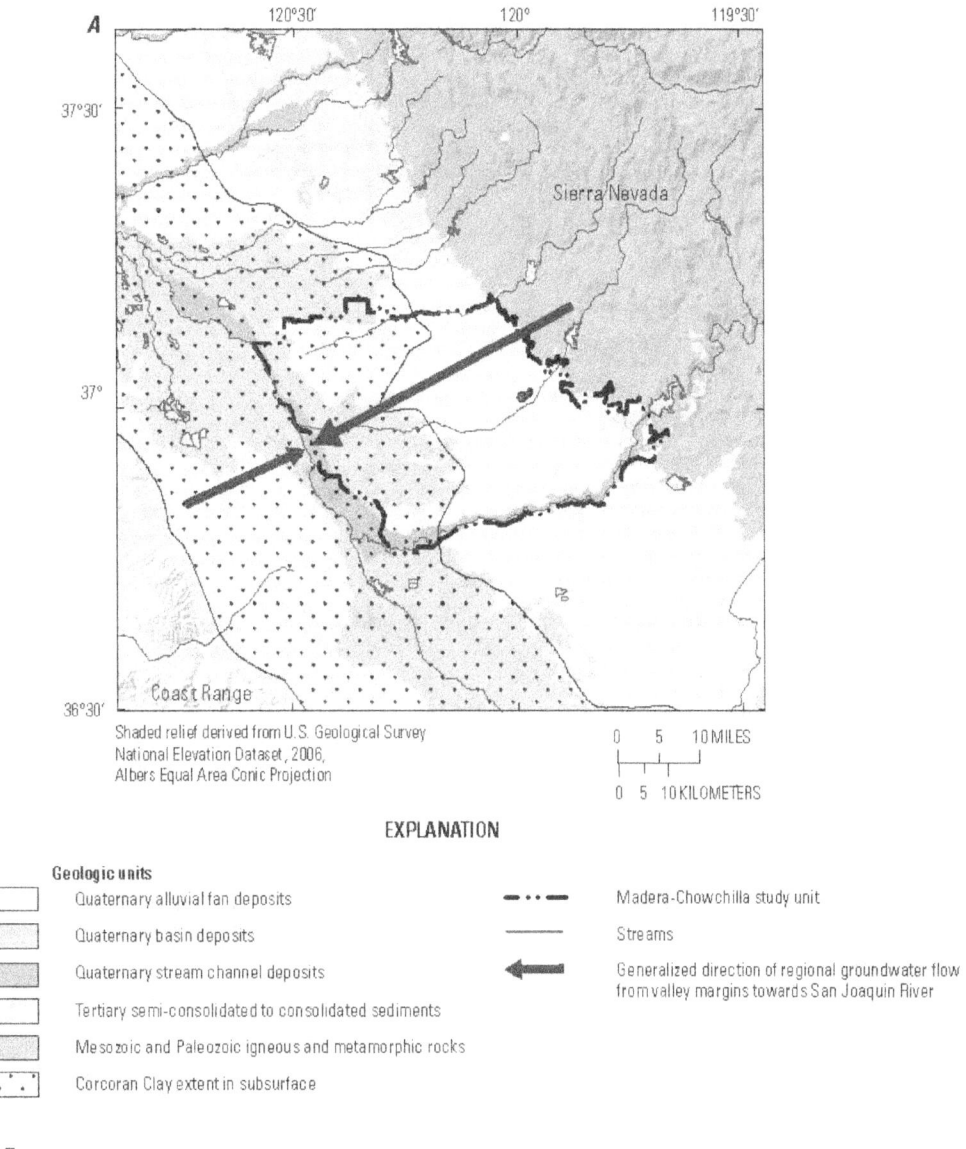

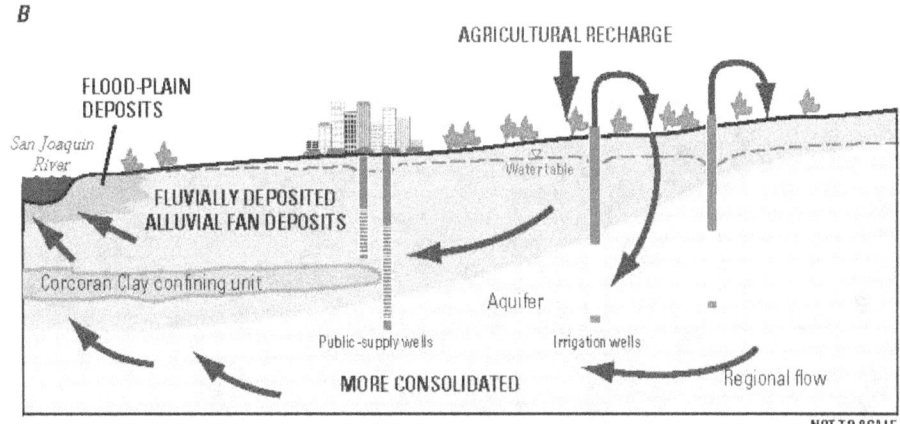

Figure 4.   (*A*) Regional lateral groundwater flow and (*B*) vertical flow influenced by agricultural practices and natural discharge zones for the aquifer system of the Madera-Chowchilla study unit, California GAMA Priority Basin Project (modified from Burow and others, 2004; Phillips and others, 2007).

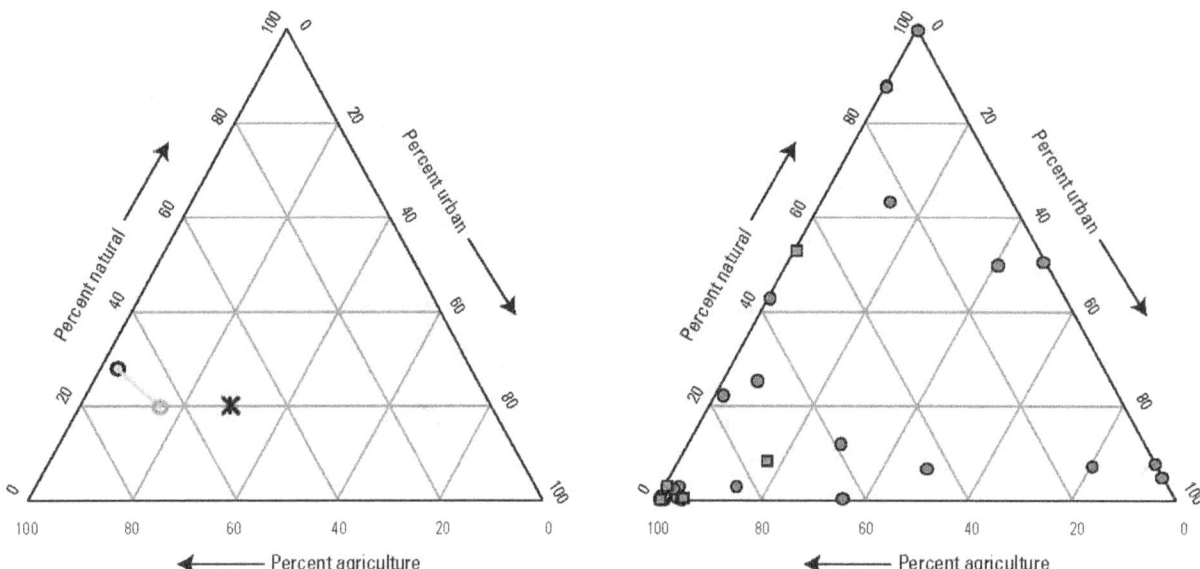

A. Whole study unit and study unit based on 500-meter buffers surrounding grid and CDPH wells.

B. Grid and understanding wells in study unit based on land use within a 500-meter buffer surrounding each well.

EXPLANATION

Study unit land use

⊙ Average for whole study unit

⊙ Average in 500-meter buffers surrounding USGS grid wells

✖ Average in 500-meter buffer surrounding CDPH wells with water quality data in 3-year period 2/12/05–2/12/08

Wells

⬤ Grid

▣ Understanding

Figure 5. Percentages of urban, agricultural, and natural land uses for (*A*) the study unit and (*B*) the areas surrounding individual wells, Madera-Chowchilla study unit, 2008, California GAMA Priority Basin Project.

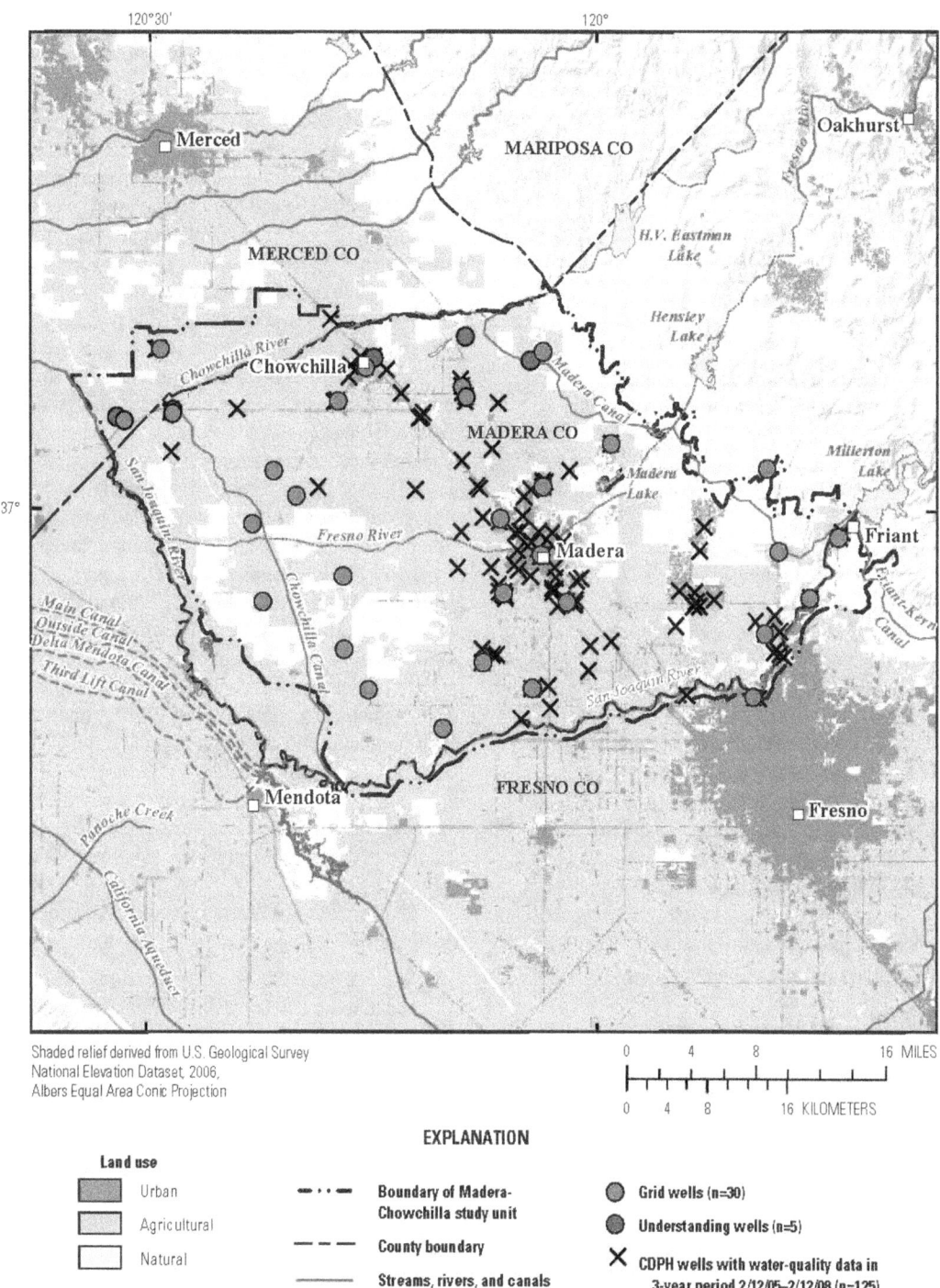

Figure 6. Land use of the Madera-Chowchilla study unit and locations of grid and understanding wells sampled during April–May 2008, and wells with data in the California Department of Public Health database during February 2005–February 2008, Madera-Chowchilla study unit, California GAMA Priority Basin Project.

Methods

The methods described here are those used for the status and understanding assessments. Methods used to collect and analyze groundwater samples and results for quality-control assessment for the constituents listed in table 1 are described by Shelton and others (2009). Methods used for compilation of data on potential explanatory factors are described in appendix A.

Status Assessment

The *status assessment* is designed to quantify groundwater quality in areal proportions of the primary aquifer system. This section describes the methods used for (1) defining groundwater quality, (2) assembling the datasets used for the assessment, (3) determining which constituents warrant additional assessment, and (4) calculating aquifer-scale proportions.

Groundwater quality was defined in terms of relative-concentration (RC), which compares the concentrations of constituents in groundwater to the concentrations of regulatory and non-regulatory benchmarks used to evaluate drinking-water quality. Constituents were selected for additional evaluation in the *status assessment* on the basis of objective criteria by using these RCs. Groundwater-quality data collected by the USGS for the GAMA Priority Basin Project (USGS-GAMA) and data compiled from the CDPH database were used in the *status assessment*. Two statistical methods were used to calculate the areal proportions of the primary aquifer system occupied by groundwater with high, moderate, or low RCs for a constituent or constituent class (aquifer-scale proportions): (1) the "grid-based" method, which uses one value per grid cell to represent groundwater quality (Belitz and others, 2010), and (2) the "spatially weighted" method, which uses many values per grid cell.

The CDPH database contains historical records from more than 25,000 wells, necessitating targeted retrievals to effectively access water-quality data. For example, for the area representing the Madera-Chowchilla study unit, the historical CDPH database contains more than 60,000 records from 154 wells. The CDPH data were used in two ways in the *status assessment*: (1) to help identify constituents for additional evaluation in the assessment, and (2) to provide the majority of the data used in the spatially weighted calculations of aquifer-scale proportions.

Relative-Concentrations and Water-Quality Benchmarks

Relative-concentrations were used to provide context for water-quality data by comparing measured concentrations of constituents in groundwater samples to water-quality benchmarks that are generally applied to finished drinking water:

$$Relative\ concentration\ (RC) = \frac{Sample\ concentration}{Benchmark\ concentration}$$

An RC less than 1 ($<$1.0) indicates a sample concentration less than the benchmark, and an RC greater than 1 ($>$1.0) indicates a sample concentration greater than the benchmark. The use of RCs also permits comparison of constituents present at a wide range of concentrations on a single scale.

Toccalino and others (2004), Toccalino and Norman (2006), and Rowe and others (2007) previously used RC by converting concentration to Benchmark Quotient (BQ), which is the ratio of measured concentration to a water-quality benchmark. The BQ concept is used in this study; however, the ratio of measured concentration to a water-quality benchmark is called relative-concentration (RC) rather than a BQ because the benchmarks used may be different. BQs are calculated using U.S. Environmental Protection Agency (USEPA) maximum contaminant levels (MCL-US) or USGS-USEPA Health-Based Screening Levels (HBSLs). HBSLs are determined using USEPA methodologies for establishing drinking-water guidelines and the most recent USEPA peer-reviewed, publically available human-health toxicity information (Toccalino, 2007). RCs are calculated using benchmarks established by USEPA and CDPH (U.S. Environmental Protection Agency, 2008a,b; California Department of Public Health, 2008a,b). HBSLs were not used in this study because they are not recognized by California drinking-water regulatory agencies, and the GAMA Program was intended to specifically focus on groundwater quality in California.

The benchmarks used to calculate RCs for each constituent were selected in the following order of priority:

1. Regulatory, health-based CDPH and USEPA maximum contaminant levels (MCL-CA and MCL-US, respectively) and action levels (AL-US).

2. Non-regulatory CDPH and USEPA secondary maximum contaminant levels (SMCL-CA and SMCL-US, respectively). For constituents with recommended and upper SMCL-CA levels, the values for the upper levels were used.

3. Non-regulatory, health-based CDPH notification levels (NL-CA), USEPA lifetime health advisory levels (HAL-US), and USEPA risk-specific doses. Risk-specific doses for risks of 1 in 10^5 (RSD5-US) were calculated by dividing the USEPA values for the concentration of a constituent in drinking water corresponding to an estimated excess lifetime cancer risk of 1 in 10^4 by 10.

Note that for constituents with multiple types of benchmarks, this hierarchy may not result in selection of the benchmark with the lowest concentration. Additional information on the types of benchmarks and the benchmarks for all constituents analyzed is provided by Shelton and others (2009).

For ease of discussion, RCs of constituents were classified into low, moderate, and high categories:

Category	Organic and special-interest constituents	Inorganic constituents
High	RC > 1.0	RC > 1.0
Moderate	$1.0 \geq RC > 0.1$	$1.0 \geq RC > 0.5$
Low	$0.1 \geq RC$	$0.5 \geq RC$

The boundary between "moderate" and "low" RCs was set at 0.1 for organic and special-interest constituents for consistency with other studies and reporting requirements (U.S. Environmental Protection Agency, 1998; Toccalino, 2007). For organic constituents, detection at concentrations greater than one-tenth of a health-based benchmark value (RC>0.1) commonly is used to identify constituents that may warrant additional monitoring to evaluate trends in their occurrences. Organic constituents generally are human-made, are—ideally—not present in groundwater, and are infrequently detected at RCs greater than 0.1. Of the three special-interest constituents, two are organic compounds [1,2,3-trichloropropane (1,2,3-TCP) and *N*-nitrosodimethylamine (NDMA)]. The third, perchlorate, is an inorganic compound and is in the special-interest group, rather than the inorganic constituent group, because at the inception of the GAMA Priority Basin Project, the State of California was assessing potential regulation of perchlorate concentrations in drinking water and therefore had a "special interest" in perchlorate occurrence. An MCL-CA was promulgated in October 2007.

For inorganic constituents, the boundary between "moderate" and "low" RCs was set at 0.5. The primary reason for using a higher boundary value was to focus attention on the inorganic constituents of most immediate concern (Fram and Belitz, 2012). The naturally occurring inorganic constituents tend to be more prevalent than organic constituents in groundwater. While more complex classifications could be devised based upon the properties and sources of individual constituents, use of a single moderate/low boundary value for each of the two major groups of constituents provided a consistent objective criteria for distinguishing constituents occurring at moderate rather than low concentrations.

Datasets for Status Assessment

Two datasets were used in the status assessment: data from grid wells, and a combination of data from grid wells, understanding wells, and CDPH wells. This section describes how each dataset was constructed. Comparisons of USGS-GAMA and CDPH data are presented in appendix B.

Grid Wells

The grid-based calculations of aquifer-scale proportions used data from 30 wells sampled by the USGS for spatial coverage of one well per grid cell across the study unit (grid wells). Detailed descriptions of the methods used to identify wells for sampling are given in Shelton and others (2009). Briefly, the Madera-Chowchilla study unit area was subdivided into 30 equal-area grid cells of 30 square miles (mi²) (fig. 2), and in each cell, one well was randomly selected to represent the cell (Scott, 1990). Wells primarily were selected from the population of wells in the statewide database maintained by the CDPH (Shelton and others, 2009). Of the 30 grid wells, 19 were listed in the CDPH database; of the remaining 11 wells, 6 were irrigation wells, and 5 were domestic wells. The depths of the perforated or open intervals in the irrigation and domestic wells were similar to the depths of the screened intervals in CDPH wells in the study unit (appendix B). The wells were numbered in the order of sample collection with the prefix "MADCHOW" identifying the study unit (fig. 2; Shelton and others, 2009; appendix A).

Samples were collected by the USGS from the grid wells and were analyzed for 282 to 288 constituents (table 1). All wells were sampled for alkalinity, dissolved oxygen, pH, specific conductance, temperature, volatile organic compounds (VOCs), pesticides, perchlorate, low-level 1,2,3-TCP, nutrients, major and minor ions, trace elements, noble gases, tritium, stable isotopes of water, uranium isotopes, carbon isotopes, and gross alpha and beta particle activity. Samples for analysis of turbidity, NDMA, pharmaceuticals, and arsenic and iron species were collected from some wells. Of the constituents with water-quality benchmarks, only NDMA was not analyzed in samples from all wells. The collection, analysis, and quality assurance of the constituents listed in table 1 are described in Shelton and others (2009). Pharmaceutical compounds were not detected at concentrations greater than or equal to method detection limits in the Madera-Chowchilla study unit. Fram and Belitz (2011a) present all results for pharmaceutical compounds in groundwater samples collected during May 2004 through March 2010 for 28 GAMA Priority Basin Project study units.

Additional Data Used for Spatially Weighted Calculation

The spatially weighted calculations of aquifer-scale proportions used USGS-GAMA data from the 30 grid and the 5 understanding wells, and data from CDPH wells reported from the most recent 3 years available in the CDPH database at the time of the USGS sampling for the GAMA Program.

Table 1.   Summary of number of wells and constituents by sampling schedule and analyte group, Madera-Chowchilla study unit, 2008, California GAMA Priority Basin Project.

[**Sampling schedule**: "Intermediate" and "slow" sampling schedules refer to the amount of time required for a field crew to complete all work at a well. Typically, two intermediate wells or one intermediate well and one slow well could be sampled in one day. **Abbreviations:** GAMA, Groundwater Ambient Monitoring and Assessment Program; LRL, laboratory reporting level]

	Sampling schedule	
	Intermediate	Slow
Wells	**Number of wells**	
Total number of wells	25	10
Number of grid wells sampled	25	5
Number of understanding wells sampled	0	5
Analyte groups	**Number of constituents**	
Inorganic constituents		
Nutrients	5	5
Major and minor ions and trace elements	36	36
Uranium isotopes[1]	1	1
Gross alpha and beta particle activity[2]	2	2
Specific conductance and alkalinity	2	2
Organic and special-interest constituents[3]		
Volatile organic compounds (VOCs) (includes fumigants)	85	85
Pesticides and degradates[4]	79	79
Polar pesticides and degradates[5]	55	55
Perchlorate and low-level 1,2,3-trichloropropane (1,2,3-TCP)[6]	2	2
N-nitrosodimethylamine (NDMA)	0	1
Geochemical and age-dating tracers		
Dissolved oxygen, pH, and temperature	3	3
Noble gases and tritium[7]	7	7
Stable isotopes of water	2	2
Tritium[8]	1	1
Carbon isotopes	2	2
Turbidity	0	1
Arsenic and iron species	0	4
Total number of constituents analyzed by sampling schedule	282	288

[1] Three uranium isotopes were measured: uranium-234, uranium-235, and uranium-238. Uranium isotope samples were not collected at five USGS-grid wells sampled on the intermediate schedule.

[2] Both 72-hour and 30-day counts were measured for alpha and beta particle activities.

[3] Twelve pharmaceutical compounds were analyzed in samples from slow wells. Because pharmaceuticals are not discussed in this report, they are not included in the count of constituents analyzed.

[4] Does not include two constituents in common with polar pesticides and degradates (carbofuran and metalaxyl).

[5] Does not include three constituents in common with pesticides and degradates (atrazine, deethylatrazine, and tebuthiuron).

[6] Includes one analyte, 1,2,3-TCP, in common with VOC analyses. The LRL for the low-level analysis is 0.0050 microgram per liter (µg/L) compared to 0.12 µg/L for the VOC analysis; therefore, the low-level analysis is counted as a separate analysis.

[7] Analyzed at Lawrence Livermore National Laboratory, Livermore, California.

[8] Analyzed at USGS Stable Isotope and Tritium Laboratory, Menlo Park, California.

In addition to the 30 grid wells, 5 understanding wells in the Madera-Chowchilla study unit were sampled by USGS-GAMA. The understanding wells were located along the northern margin of the study unit and were selected to help identify differences in water quality with depth in the primary aquifer system. The understanding wells were numbered in the order of collection with the prefix "MADCHOWFP" (fig. 2; Shelton and others, 2009; appendix A).

The CDPH database contained water-quality data for 154 wells. Of these 154 wells, 125 had water-quality data for samples collected between February 12, 2005, and February 12, 2008 (fig. 2), the most recent 3-year interval of data available from the CDPH database at the time of USGS-GAMA sampling in the study unit. These 125 wells provided the bulk of the data for the spatially weighted calculations. For wells with multiple analyses for a constituent during the 3-year interval, the most recent analysis was used. For the 20 wells (19 grid wells and 1 understanding well) with CDPH and USGS data, only the USGS data were used.

Reporting limits for inorganic constituents in the CDPH database were at concentrations below RCs of 0.5 for all constituents except antimony and thallium. Constituent concentration data from USGS-GAMA analysis and the CDPH database therefore can be adequately classified as having high, moderate, or low RCs. For organic (VOCs and pesticides) and special-interest constituents (NDMA and perchlorate), however, reporting limits used by USGS-GAMA were significantly lower than those in the CDPH database (table 2), and CDPH reporting limits had concentrations greater than RCs of 0.5 for some constituents. In addition, USGS-GAMA analyzed many more constituents than are reported in the CDPH database (table 2); therefore, the spatially weighted calculations of aquifer-scale proportions may provide only minimum estimates of the proportion of moderate values for those constituents.

Identification of Constituents for Additional Evaluation in the Status Assessment

Up to 288 constituents were analyzed by USGS-GAMA in samples from wells in the Madera-Chowchilla as part of the *status assessment* (table 1); however, only a subset of these constituents was selected for additional evaluation in this report. Three criteria were used to select constituents for additional evaluation:

1. Constituents present at high or moderate RCs in the CDPH database within the 3-year interval (February 12, 2005–February 12, 2008) prior to the USGS-GAMA sampling period,

2. Constituents present at high or moderate RCs in the grid wells or understanding wells used in the status assessment, or

3. Organic constituents having study-unit detection frequencies greater than or equal to 10% in the grid well dataset regardless of concentration.

These criteria identified 14 inorganic constituents and 10 organic constituents and special-interest constituents for additional evaluation in the *status assessment* (table 3). An additional 30 inorganic constituents and 16 organic constituents and all 20 geochemical and age-dating tracers were detected in the wells sampled by the USGS in April and May 2008, but were not selected for additional evaluation in the *status assessment* because they either have no established benchmarks, or they were only detected at low RCs and, for organic constituents, had study unit detection frequencies less than 10% (table 4). The remaining 198 constituents that were analyzed but not detected by USGS-GAMA are listed in Shelton and others (2009).

Table 2. Comparison of number of constituents analyzed and median method detection levels or laboratory reporting levels by type of constituent for data reported in the California Department of Public Health database and for data collected by the U.S. Geological Survey (USGS) for the Madera-Chowchilla study unit, 2008, California GAMA Priority Basin Project.

[**Abbreviations:** GAMA, Groundwater Ambient Monitoring and Assessment Program; CDPH, California Department of Public Health; MDL, method detection level; LRL, laboratory reporting level; µg/L, micrograms per liter; NDMA, *N*-nitrosodimethylamine]

Constituent type	CDPH		USGS-GAMA		Median units
	Number of constituents	Median MDL	Number of constituents	Median LRL	
Volatile organic compounds (including fumigants)	65	0.50	85	0.06	µg/L
Pesticides and degradates, and polar pesticides and metabolites	43	1	134	0.020	µg/L
NDMA	1	5	1	0.0020	µg/L
Perchlorate	1	4	1	0.10	µg/L

Table 3.   Benchmark type and value for constituents selected for additional evaluation in the status assessment of groundwater quality in the Madera-Chowchilla study unit, 2008, California GAMA Priority Basin Project.

[**Benchmark type:** AL-US, USEPA action level; HAL-US, USEPA lifetime health advisory level; MCL-CA, CDPH maximum contaminant level; MCL-US, USEPA maximum contaminant level; NL-CA, CDPH notification level; RSD5-US, USEPA risk-specific dose at a risk factor of 10^{-5}; SMCL-CA, CDPH secondary maximum contaminant level. **Benchmark units:** μg/L, micrograms per liter; mg/L, milligrams per liter; μS/cm at 25°C, microsiemens per centimeter at 25 degrees Celsius; pCi/L, picocuries per liter. **Other abbreviations:** GAMA, Groundwater Ambient Monitoring and Assessment Program; USEPA, U.S. Environmental Protection Agency; CDPH, California Department of Public Health. **Constituent names:** DBCP, 1,2-dibromo-3-chloropropane; EDB, 1,2-dibromoethane; NDMA, *N*-nitrosodimethylamine; PCE, tetrachloroethene; 1,2,3-TCP, 1,2,3-trichloropropane; TDS, total dissolved solids; THM, trihalomethane]

Constituent	Typical use or source	Benchmark type	Benchmark value	Benchmark units
Organic constituents with health-based benchmarks				
Atrazine	Herbicide	MCL-CA	1	μg/L
Diuron	Herbicide	RSD5-US	20	μg/L
Simazine	Herbicide	MCL-US	4	μg/L
DBCP	Fumigant	MCL-US	0.2	μg/L
EDB	Fumigant	MCL-US	0.05	μg/L
1,2,3-TCP[1]	Fumigant	HAL-US	40	μg/L
PCE	Solvent	MCL-US	5	μg/L
Chloroform[2]	Disinfection byproduct (THM)	MCL-US	80	μg/L
Constituents of special interest				
NDMA	Rocket fuel, disinfection byproduct	NL-CA	0.01	μg/L
Perchlorate	Natural, rocket fuel, fireworks	MCL-CA	6	μg/L
Inorganic constituents with health-based benchmarks				
Nutrients				
Nitrate plus nitrite, as nitrogen	Natural, fertilizer, sewage	MCL-US	10	mg/L
Trace elements				
Arsenic	Naturally occurring	MCL-US	10	μg/L
Barium	Naturally occurring	MCL-CA	1,000	μg/L
Lead	Naturally occurring	AL-US	15	μg/L
Strontium	Naturally occurring	HAL-US	4,000	μg/L
Uranium	Naturally occurring	MCL-US	30	μg/L
Vanadium	Naturally occurring	NL-CA	50	μg/L
Radioactive constituents				
Gross alpha particle activity	Naturally occurring	MCL-US	15	pCi/L
Uranium activity	Naturally occurring	MCL-CA	20	pCi/L
Inorganic constituents with aesthetic or technical (SMCL) benchmarks				
Chloride	Naturally occurring	SMCL-CA	500	mg/L
TDS	Naturally occurring	SMCL-CA	1,000	mg/L
Specific conductance	Naturally occurring	SMCL-CA	1,600	μS/cm at 25°C
Iron	Naturally occurring	SMCL-CA	300	μg/L
Manganese	Naturally occurring	SMCL-CA	50	μg/L

[1] HAL-US was eliminated as of October 2009. NL-CA for 1,2,3-TCP is 0.005 μg/L. This report uses the HAL-US instead of NL-CA as the benchmark for 1,2,3-TCP because the NL-CA is less than the lowest reporting limit available for 1,2,3-TCP.

[2] Chloroform is a trihalomethane (THM). The MCL-US benchmark is the sum of chloroform, bromoform, bromodichloromethane, and dibromochloromethane.

Table 4. Constituents detected in samples collected, but not selected for additional evaluation in the status assessment for the Madera-Chowchilla study unit, 2008, California GAMA Priority Basin Project.

[**Benchmark types:** MCL-US, USEPA maximum contaminant level; MCL-CA, CDPH maximum contaminant level; AL-US, USEPA action level; HAL-US, USEPA lifetime health advisory level; NL-CA, CDPH notification level; SMCL-US, USEPA secondary maximum contaminant level; SMCL-CA, CDPH secondary maximum contaminant level. **Other abbreviations:** GAMA, Groundwater Ambient Monitoring and Assessment Program; USEPA, U.S. Environmental Protection Agency; CDPH, California Department of Public Health]

Constituent	Typical use or source	Benchmark type
Inorganic constituents with health-based benchmarks		
Aluminum	Naturally occurring	MCL-CA
Ammonia (as nitrogen)	Naturally occurring	HAL-US
Antimony	Naturally occurring	MCL-US
Beryllium	Naturally occurring	MCL-US
Boron	Naturally occurring	NL-CA
Cadmium	Naturally occurring	MCL-US
Chromium	Naturally occurring	MCL-CA
Copper	Naturally occurring	AL-US
Fluoride	Naturally occurring	MCL-CA
Gross beta particle activity	Naturally occurring	MCL-CA
Molybdenum	Naturally occurring	HAL-US
Nickel	Naturally occurring	MCL-CA
Nitrite (as nitrogen)	Naturally occurring	MCL-US
Selenium	Naturally occurring	MCL-US
Inorganic constituents with aesthetic/technical-based benchmarks		
Silver	Naturally occurring	SMCL-CA
Sulfate	Naturally occurring	SMCL-CA
Zinc	Naturally occurring	SMCL-CA
Inorganic constituents with no benchmarks		
Alkalinity	Naturally occurring	None
Bromide	Naturally occurring	None
Calcium	Naturally occurring	None
Cobalt	Naturally occurring	None
Iodide	Naturally occurring	None
Lithium	Naturally occurring	None
Magnesium	Naturally occurring	None
Nitrogen, total	Naturally occurring	None
Orthophosphate, as phosphorus	Naturally occurring	None
Potassium	Naturally occurring	None
Silica	Naturally occurring	None
Sodium	Naturally occurring	None
Tungsten	Naturally occurring	None
Organic constituents with regulatory, health-based benchmarks		
1,1-Dichloroethane (1,1-DCA)	Solvent	MCL-CA
1,2-Dichloropropane	Fumigant	MCL-US
Bromodichloromethane	Disinfection byproduct (THM)	MCL-US
Bromoform (Tribromomethane)	Disinfection byproduct (THM)	MCL-US
cis-1,2-Dichloroethene (cis-1,2-DCE)	Solvent	MCL-CA
Dibromochloromethane	Disinfection byproduct (THM)	MCL-US
Dinoseb (Dinitrobutyl phenol)	Herbicide	MCL-CA
Trichloroethene (TCE)	Solvent	MCL-US

Table 4.   Constituents detected in samples collected, but not selected for additional evaluation in the status assessment for the Madera-Chowchilla study unit, 2008, California GAMA Priority Basin Project.—Continued

[**Benchmark types:** MCL-US, USEPA maximum contaminant level; MCL-CA, CDPH maximum contaminant level; AL-US, USEPA action level; HAL-US, USEPA lifetime health advisory level; NL-CA, CDPH notification level; SMCL-US, USEPA secondary maximum contaminant level; SMCL-CA, CDPH secondary maximum contaminant level. **Other abbreviations:** GAMA, Groundwater Ambient Monitoring and Assessment Program; USEPA, U.S. Environmental Protection Agency; CDPH, California Department of Public Health]

Constituent	Typical use or source	Benchmark type
Organic constituents with non-regulatory, health-based benchmarks		
Bromacil	Herbicide	HAL-US
Hexazinone	Herbicide	HAL-US
Tebuthiuron	Herbicide	HAL-US
Organic constituents with no benchmarks		
3,4-Dichloroaniline	Herbicide degradate (diuron)	None
Deethylatrazine (2-Chloro-4-isopropylamino-6-amino-s-triazine; DEA)	Herbicide degradate (atrazine)	None
Deisopropyl atrazine (2-Chloro-6-ethylamino-4-amino-s-triazine; DIA)	Herbicide degradate (atrazine)	None
Imazethapyr	Herbicide	None
Norflurazon	Herbicide	None
Geochemical and age-dating tracers		
Tritium	Naturally occurring	MCL-CA
pH	Naturally occurring	SMCL-US
Dissolved oxygen, temperature	Naturally occurring	None
Turbidity	Naturally occurring	None
Carbon-14 and $\delta^{13}C$ of dissolved carbonates	Naturally occurring	None
δ^2H and $\delta^{18}O$ stable isotopes of water	Naturally occurring	None
Five noble gases, tritium, and δ^3He	Naturally occurring	None
Four arsenic and iron species	Naturally occurring	None

Table 5.   Constituents historically reported at high relative-concentrations in the California Department of Public Health database for the Madera-Chowchilla study unit, 2008, California GAMA Priority Basin Project.

[The historical period of CDPH well data is from January 6, 1984, to February 11, 2005. The 3-year period used in the status assessment is from February 12, 2005, to February 12, 2008. Relative-concentration equals measured concentration divided by benchmark value; relative-concentration greater than 1 is defined as high. **Benchmark types:** AL-US, USEPA action level; MCL-US, USEPA maximum contaminant level; MCL-CA, CDPH maximum contaminant level. **Benchmark units:** µg/L, micrograms per liter; mg/L, milligrams per liter; pCi/L, picocuries per liter. **Other abbreviations:** CDPH, California Department of Public Health; GAMA, Groundwater Ambient Monitoring and Assessment Program; USEPA, U.S. Environmental Protection Agency]

Constituent	Benchmark type	Benchmark value	Units	Date of most recent high value	Number of historically high wells	Number of wells with analysis
Cadmium	MCL-US	5	µg/L	02-15-98	1	84
EDB[1,2]	MCL-US	0.05	µg/L	12-11-07	1	101
Fluoride[1]	MCL-CA	2	mg/L	01-24-06	3	93
Lead[1,2]	AL-US	15	µg/L	06-27-06	6	91
Mercury	MCL-US	2	µg/L	04-19-85	2	84
Radium[1]	MCL-US	5	pCi/L	08-18-05	1	73

[1]1,2-Dibromoethane (EDB), fluoride, lead, and radium were reported at high relative-concentrations between February 12, 2005, and February 12, 2008, but the high value was not the most recent value reported for the well.

[2]1,2-Dibromoethane (EDB) was detected at moderate relative-concentration in USGS-GAMA samples, and lead was reported at moderate relative-concentration in the CDPH database between February 12, 2005, and February 12, 2008, thus, both constituents met the criteria for additional evaluation in the status assessment and are listed in table 8.

The USGS conducted a review of the water-quality data (January 6, 1984, to February 12, 2008) in the CDPH database to identify constituents that have been reported at high RCs historically, but not currently. Constituents may be historically high, but not currently high, because of improvement of groundwater quality with time or abandonment of wells with high concentrations. Constituents with historically high RCs that do not otherwise meet the criteria for selection for additional evaluation in the *status assessment* are not considered representative of current potential groundwater-quality concerns in the study unit. Two constituents, cadmium and mercury, were reported at high RCs before the 3-year interval used for the *status assessment* (table 5). Cadmium was detected at low concentrations in several samples, and mercury was not detected in any samples analyzed by USGS-GAMA. Four constituents, 1,2-dibromoethane, fluoride, lead, and radium, were reported at high RCs during the 3-year interval, but the most recent sample used in the *status assessment* did not have high RCs (table 5).

Calculation of Aquifer-Scale Proportions

The *status assessment* is intended to characterize the current quality of groundwater resources within the primary aquifer system of the Madera-Chowchilla study unit. The primary aquifer system is defined by the depth intervals over which wells listed in the CDPH database are perforated. The use of the term "primary aquifer system" does not imply that there exists a discrete aquifer unit. In most groundwater basins, public supply wells typically are perforated at greater depths than are domestic wells (Burow and others, 2008). Thus, because domestic wells are not listed in the CDPH database, the primary aquifer system generally corresponds to the portion of the aquifer system tapped by public wells. However, to the extent that domestic wells in the study unit are perforated over the same depth intervals as the CDPH wells, the assessments presented in this report may also be applicable to the portions of the aquifer systems used for domestic drinking-water supplies (appendix B).

Two statistical approaches, grid-based and spatially weighted, were selected to evaluate the proportions of the primary aquifer system with high and moderate RCs of constituents (Belitz and others, 2010). For ease of discussion, these proportions are referred to as "high" and "moderate" aquifer-scale proportions. Calculations of aquifer-scale proportions were made for individual constituents meeting the criteria for additional evaluation in the *status assessment*, and for classes of constituents. Classes of constituents with health-based benchmarks included nutrients, trace elements, radioactive constituents, herbicides, fumigants, solvents, and THMs. Class of constituents with aesthetic-based benchmarks included salinity indicators and trace elements.

The grid-based calculation uses the grid-well dataset. For each constituent, the high aquifer-scale proportion was calculated by dividing the number of cells represented by a high value for that constituent by the total number of grid cells with data for that constituent. The moderate aquifer-scale proportion was calculated similarly. Confidence intervals for the high aquifer-scale proportions were computed using the Jeffreys interval for the binomial distribution (Brown and others, 2001; Belitz and others, 2010). For calculation of high aquifer-scale proportion for a class of constituents, cells were considered high if any of the constituents had a high value. Cells were considered moderate if any of the constituents had a moderate value, but none had a high value within the cell. The grid-based estimate is spatially unbiased; however, it may not detect constituents that are present at high RCs in small proportions of the primary aquifer system.

The spatially weighted calculation uses the dataset assembled from the CDPH wells and all of the USGS wells. For each constituent, the high aquifer-scale proportion was calculated by computing the proportion of "high" wells in each cell and then averaging the proportions for all the cells (Isaaks and Srivastava, 1989; Belitz and others, 2010). The moderate aquifer-scale proportion was calculated similarly. Confidence intervals for spatially weighted detection frequencies of high concentrations are not described in this report. For calculation of high aquifer-scale proportion for a class of constituents, wells were considered high if any of the constituents had a high value. Wells were considered moderate if any of the constituents had a moderate value, but none had a high value.

In addition, for each constituent, the raw detection frequencies of high and moderate values for individual constituents were calculated using same dataset as used for the spatially weighted calculations. Raw detection frequencies are not spatially unbiased, however, because the wells in the CDPH database are not uniformly distributed (fig. 2). For example, if a constituent were present at high concentrations in a small region of the aquifer that had a high density of wells, the raw detection frequency of high values would be greater than the high aquifer-scale proportion. Raw detection frequencies are provided for reference but were not used to assess aquifer-scale proportions.

The grid-based high aquifer-scale proportions were used to represent proportions in the primary aquifer system unless the spatially weighted proportions were significantly different from the grid-based values. Significantly different results were defined as follows:

- If the grid-based high aquifer-scale proportion was zero and the spatially weighted proportion was non-zero, then the spatially weighted result was used. This situation can arise when a constituent is present at high RCs in a small proportion of the primary aquifer system.

- If the grid-based high aquifer-scale proportion was non-zero and the spatially weighted proportion was outside the 90% confidence interval (based on the Jeffreys interval for the binomial distribution), then the spatially weighted proportion was used.

The grid-based moderate and low proportions were used in most cases because the reporting limits for many organic constituents and some inorganic constituents in CDPH were higher than the boundary between the moderate and low categories. However, if the grid-based moderate proportion was zero and the spatially weighted proportion non-zero, then the spatially weighed value was used as a minimum estimate for the moderate proportion.

A subset of the constituents examined in the status assessment, as well as selected classes of constituents, was examined in the understanding assessment:

- Constituents with high aquifer-scale proportions of greater than 2%. These constituents were selected to focus the understanding assessment on those constituents that have the greatest effect on groundwater quality.

- Classes of organic constituents that included constituents detected in 10% or more of grid wells, regardless of concentration.

The *understanding assessment* was based on the 35 grid and understanding wells sampled by USGS-GAMA. CDPH wells were not used because data for many of the potential explanatory factors were not available. In particular, data for age-dating tracers, dissolved oxygen, well depth, and depth to the top of screened interval are not maintained in the CDPH database. For different potential explanatory factors, correlations were tested using either the set of grid plus understanding wells or grid wells only. Because the understanding wells were not randomly selected on a spatially distributed grid, understanding wells were excluded from analyses of relations of water quality to areally distributed variables (land use and lateral position) to avoid areal-clustering bias. Understanding wells, however, were included in analyses of relations between constituents and the vertically distributed explanatory variables depth, groundwater age, and oxidation-reduction characteristics to aid in the identification of relations.

Statistical Analysis

Nonparametric statistical methods were used to test the significance of correlations between water-quality variables and potential explanatory variables. Nonparametric statistics are robust techniques that are generally not affected by outliers and do not require that the data follow any particular distribution (Helsel and Hirsch, 2002). The significance level (p) used for hypothesis testing for this report was compared to a threshold value (α) of 5% ($\alpha = 0.05$) to evaluate whether the relation was statistically significant ($p < \alpha$). Correlations were investigated using Spearman's method to calculate the rank-order correlation coefficient (rho) between continuous variables. The values of rho can range from +1.0 (perfect positive correlation), through 0.0 (no correlation), to −1.0 (perfect negative correlation). For potential explanatory factors that were classified into categories (groundwater age, well depth, and position relative to the Corcoran Clay), the values of water-quality parameters between the categories were compared using the Wilcoxon rank-sum test. The Wilcoxon rank-sum test is a median test statistic that compares two independent data groups (categories) to determine whether one group contains larger values than the other (Helsel and Hirsch, 2002). The null hypothesis for the Wilcoxon rank-sum test is that there is no significant difference between the observations of the two independent data groups being tested. All statistical analyses were done using TIBCO Spotfire S+® 8.1 for Windows.

Potential Explanatory Factors

Brief descriptions of potential explanatory factors including land use, depth, position relative to the Corcoran Clay, lateral position, groundwater age, and geochemical conditions are given in this section. Correlations between explanatory factors that could affect apparent relations between explanatory factors and water quality also are described. The data sources and methodology used for assigning values for potential explanatory factors are described in appendix A.

Land Use

Land use based on all the land within the study unit boundaries was 69% agricultural, 28% natural, and 3% urban (fig. 5A). Compared to the land use in the entire study unit, the average land use around the CDPH wells (500-meter radius) was 26% more urban and 18% less agricultural. Average land use around the grid wells was 13% more urban and 8% less agricultural than land use in the entire study unit (fig. 5A). The difference between overall land use and land use around wells reflects that public-supply wells are often located in or near communities. The difference between the average land use around the CDPH wells and around the grid wells reflects the spatially distributed nature of the grid wells. The CDPH wells are biased towards urban land use because urbanized areas generally have a higher density of CDPH wells.

Land use surrounding two-thirds of the individual grid wells was greater than 50% agricultural (fig. 5B; table A1). Most of the remaining grid wells were surrounded by mixtures of urban and natural land use. Four of the five understanding wells had greater than 75% agricultural land use.

An additional subcategory of agricultural land use included in the analysis of fumigant concentrations was percentage of orchard/vineyard land use. Orchard/vineyard land use has previously been related to concentrations of fumigants and nitrate in parts of the eastern San Joaquin Valley (Domagalski, 1997; Burow and others, 1998a). In the Madera-Chowchilla study unit, orchard/vineyard land use occurred primarily in the south-central portion of the study unit. Land use around more than half of the individual grid wells and understanding wells was greater than 10% orchard/vineyard (table A1).

Depth, Position Relative to the Corcoran Clay, and Lateral Position

Grid wells had well depths ranging from 140 to 830 ft, with a median of 388 ft (fig. 7A; table A2). Depth to the top of the perforations ranged from 48 to 506 ft, with a median of 240 ft (fig. 7B). The perforation length ranged from 0 to 410 ft, with a median of 160 ft (fig. 7C). Three grid wells had perforation lengths of 0 ft; these wells have solid casings and draw groundwater through the open bottom of the well. The understanding wells generally were shallower with shorter perforation lengths than the grid wells (figs. 7A,C). The median well depth, depth to the top of perforations, and perforation length were 254, 212, and 42 ft, respectively, for understanding wells (figs. 7A,B,C).

The depth to the top of the Corcoran Clay dips from about 68 to 80 ft below land surface near Chowchilla to about 350 to 400 ft below land surface at the southwestern edge of the study unit (Mitten and others, 1970; Page, 1986). Most of the USGS-GAMA wells sampled (20 out of 35 wells) were located east of the extent of the Corcoran Clay. Of the 15 wells located where the Corcoran is present, 8 were perforated below it, 6 were perforated above or across it, and there was not enough well construction information for 1 well to determine the depth relative to the position of the Corcoran Clay (fig. 3; table A2).

The wells were relatively evenly distributed across the study unit between the Sierra Nevada foothills and the central axis of the San Joaquin Valley (figs. 8A,B). The central axis is defined by the reach of the San Joaquin River that flows from southeast to northwest. Lateral position is calculated as the ratio of the distance of the well from the central axis of the valley (the downgradient, or distal, end of the regional groundwater flow system) to the total distance from the central axis to the margin of the valley along the foothills (the upgradient, or proximal, end of the regional groundwater flow system). Lateral positions of grid wells range from 0.05 for the well located closest to the central axis to 1.00 for wells located closest to the foothills (table A1, figs. 8A,B).

Groundwater Age

Data for the age-dating tracers tritium, carbon-14, and helium-4 were used to classify groundwater age distributions (appendix A, tables A3 and A4). Groundwater with tritium activity greater than or equal to 0.2 tritium units (TU), terrigenic helium less than 10% of total helium, and carbon-14 greater than 80 percent modern carbon (pmc) was classified as having a "modern" age distribution (recharged since approximately 1950). Groundwater with tritium activity less than 0.2 TU and carbon-14 less than 80 pmc was classified as having a "pre-modern" age distribution (recharged before approximately 1950). Groundwater with tritium activity greater than or equal to 0.2 TU, and either terrigenic helium greater than or equal to 10% of total helium or carbon-14 less than 80 pmc was classified as having a "mixed" age distribution. Groundwater with a mixed age distribution is a mixture of waters with modern and pre-modern age distributions.

Of the 35 Madera-Chowchilla grid and understanding well samples, 9 were classified as having modern groundwater age distributions, 17 were classified as having mixed age distribution, and 9 were classified as having pre-modern age distributions (table A4). Wells yielding groundwater with pre-modern age distributions typically were deeper than wells yielding groundwater with modern age distributions and were significantly deeper than wells yielding groundwater with mixed age distributions (fig. 9A, table 6A). Wells yielding groundwater with modern age distributions had significantly shallower depths to top of perforations than wells yielding groundwater with pre-modern age distributions ($p = 0.001$) (fig. 9B, table 6A).

Classified groundwater ages and data for well depths and depths to top of perforations and bottom of perforations within the perforated interval were used to create a 3-factor classification system for well depth (fig. 10). Well depth and perforation interval information was available to classify thirty-three of the thirty-five wells sampled by USGS-GAMA as either shallow, mixed, or deep.

The boundary depth was selected to maximize the segregation of groundwater samples with modern age distributions from those with pre-modern age distributions. A boundary depth of 280 ft resulted in classification of all wells with modern age distributions as shallow or mixed depth wells, and all wells with pre-modern age distributions as deep or mixed depth wells (fig. 10).

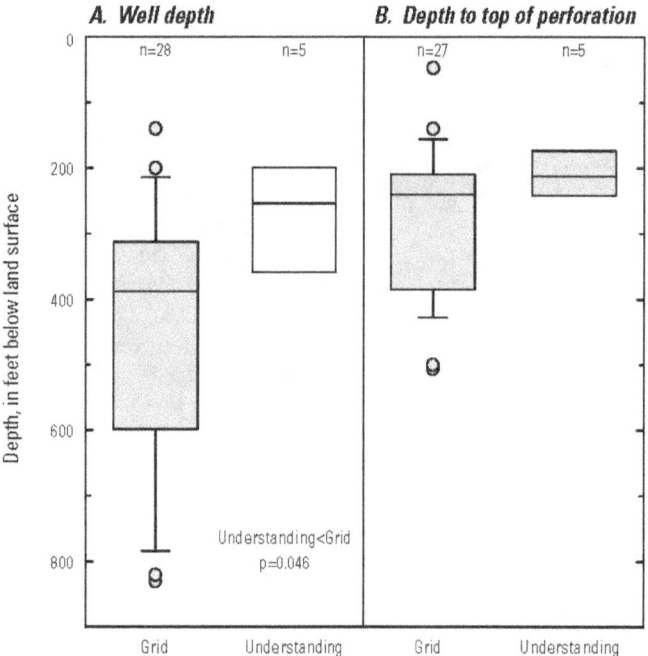

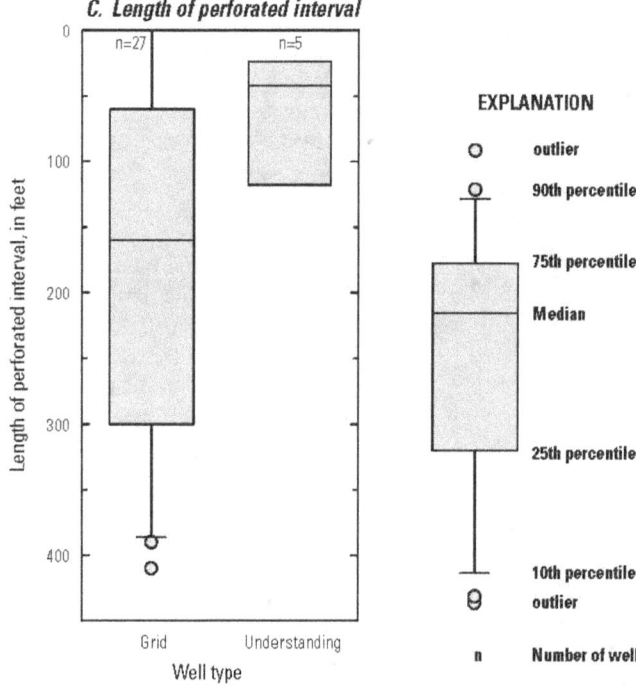

Figure 7.   (*A*) Well depth, (*B*) depth to top of perforations, and (*C*) perforation lengths for grid and understanding wells, Madera-Chowchilla study unit, 2008, California GAMA Priority Basin Project.

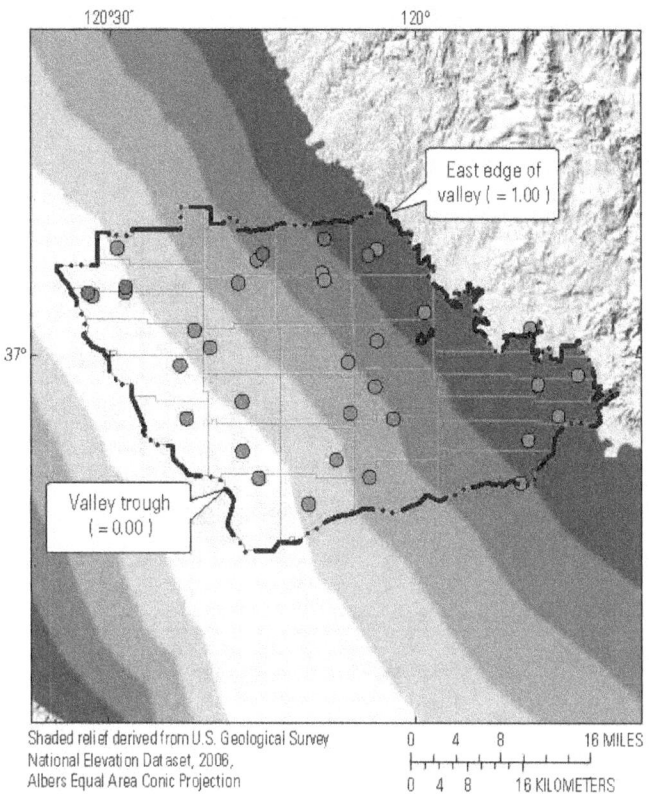

A. Distribution of wells plotted on visualization of normalized lateral positions from eastern margin of valley.

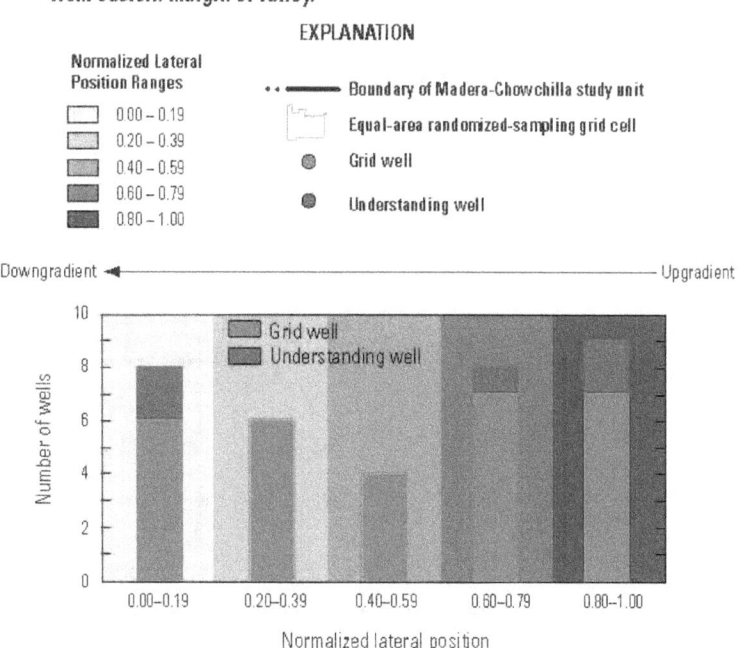

B. Distribution of wells across the range of normalized lateral position values.

Figure 8. (*A*) Distribution of wells plotted on visualization of normalized lateral positions from eastern margin of valley and (*B*) the distribution of wells across the range of normalized lateral position values, Madera-Chowchilla study unit, 2008, California GAMA Priority Basin Project.

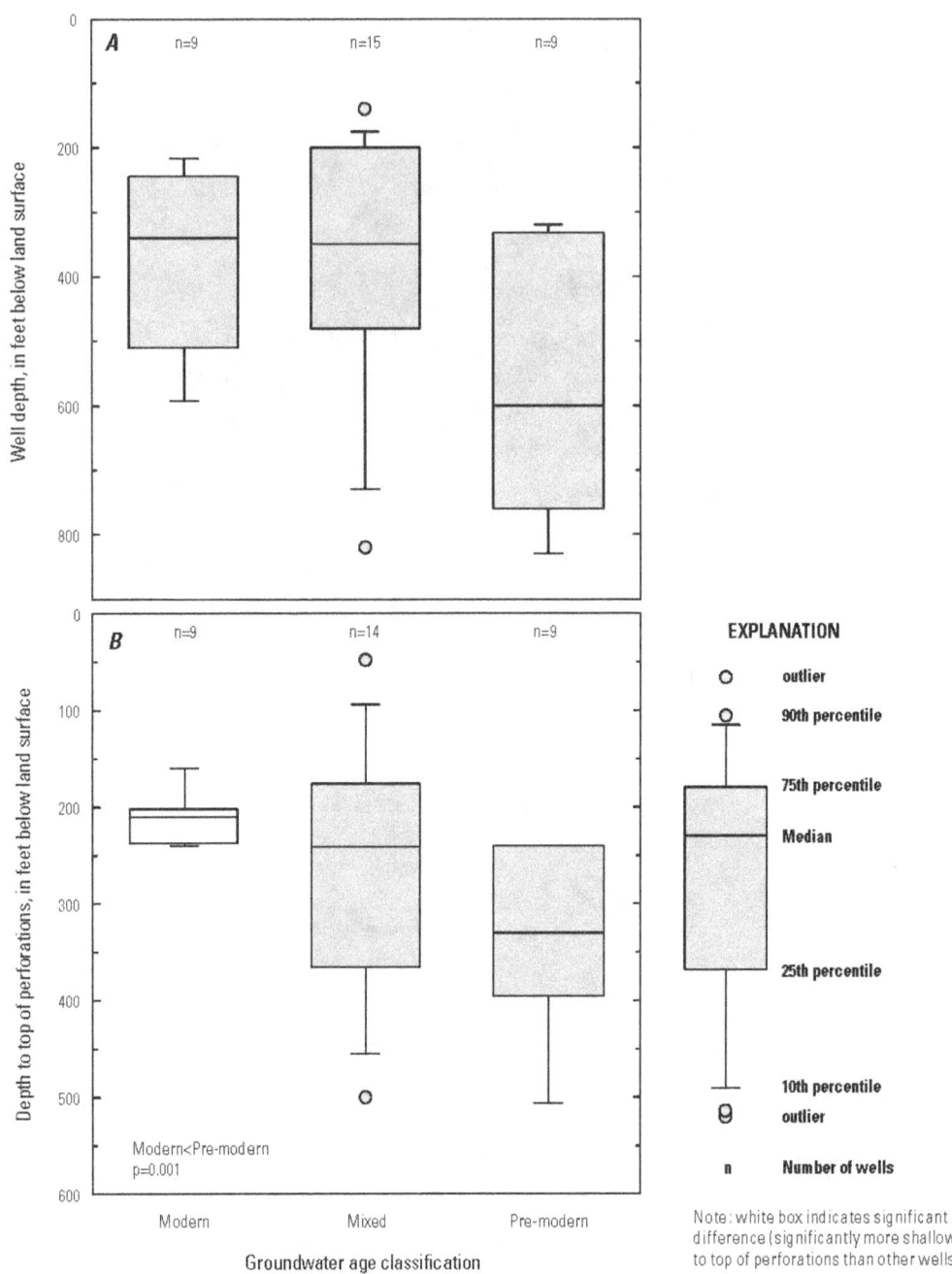

Figure 9.   Relation of classified groundwater age to (*A*) well depth and (*B*) depth to top of perforation, Madera-Chowchilla study unit, 2008, California GAMA Priority Basin Project.

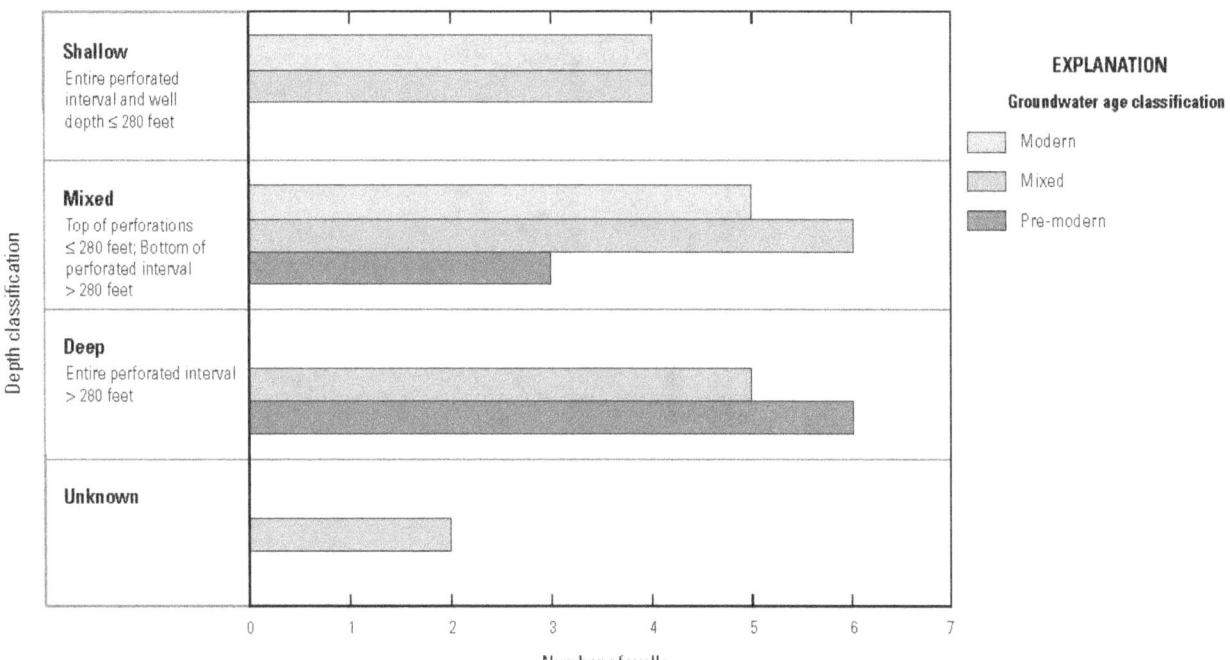

Figure 10. Groundwater age distributions in depth classes, Madera-Chowchilla study unit, 2008, California GAMA Priority Basin Project.

Geochemical Conditions

The geochemical conditions investigated as potential explanatory factors in this report are oxidation-reduction characteristics, pH, and cation ratios.

Groundwater in the Madera-Chowchilla study unit was primarily classified as oxic; 34 of the 35 wells (97%) had dissolved oxygen concentrations greater than 0.5 milligram per liter (mg/L) (table A5). The one well with anoxic conditions was located in the northwestern portion of the study unit and was entirely perforated below the Corcoran Clay (fig. 11A). All eight wells with lateral position less than 0.20 (closest to the central axis of the valley) had oxic conditions.

The pH values ranged from 6.8 to 8.4 in the 35 grid and understanding wells in the Madera-Chowchilla study unit (table A5). The highest values of pH (greater than or equal to 8) occurred in groundwater from five wells located in the northwest corner of the study unit. These five wells all had perforated intervals entirely below the Corcoran Clay (fig. 11B).

The proportions of the major cations, calcium, magnesium, sodium, and potassium, in groundwater from the study unit varied widely (table A5). The proportion is expressed as Fract-CaMg, the ratio of the sum of calcium plus magnesium to the total of the four cations in milliequivalents, with higher values indicating higher proportions of calcium plus magnesium in the cations. The highest values (0.69 to 0.79) were found in groundwater from wells with depth to top of perforations less than 235 ft and mostly with lateral positions less than 0.25 (fig. 11C). The lowest values (0.16 to 0.49) mostly were found in deeper wells, particularly wells perforated below the Corcoran Clay.

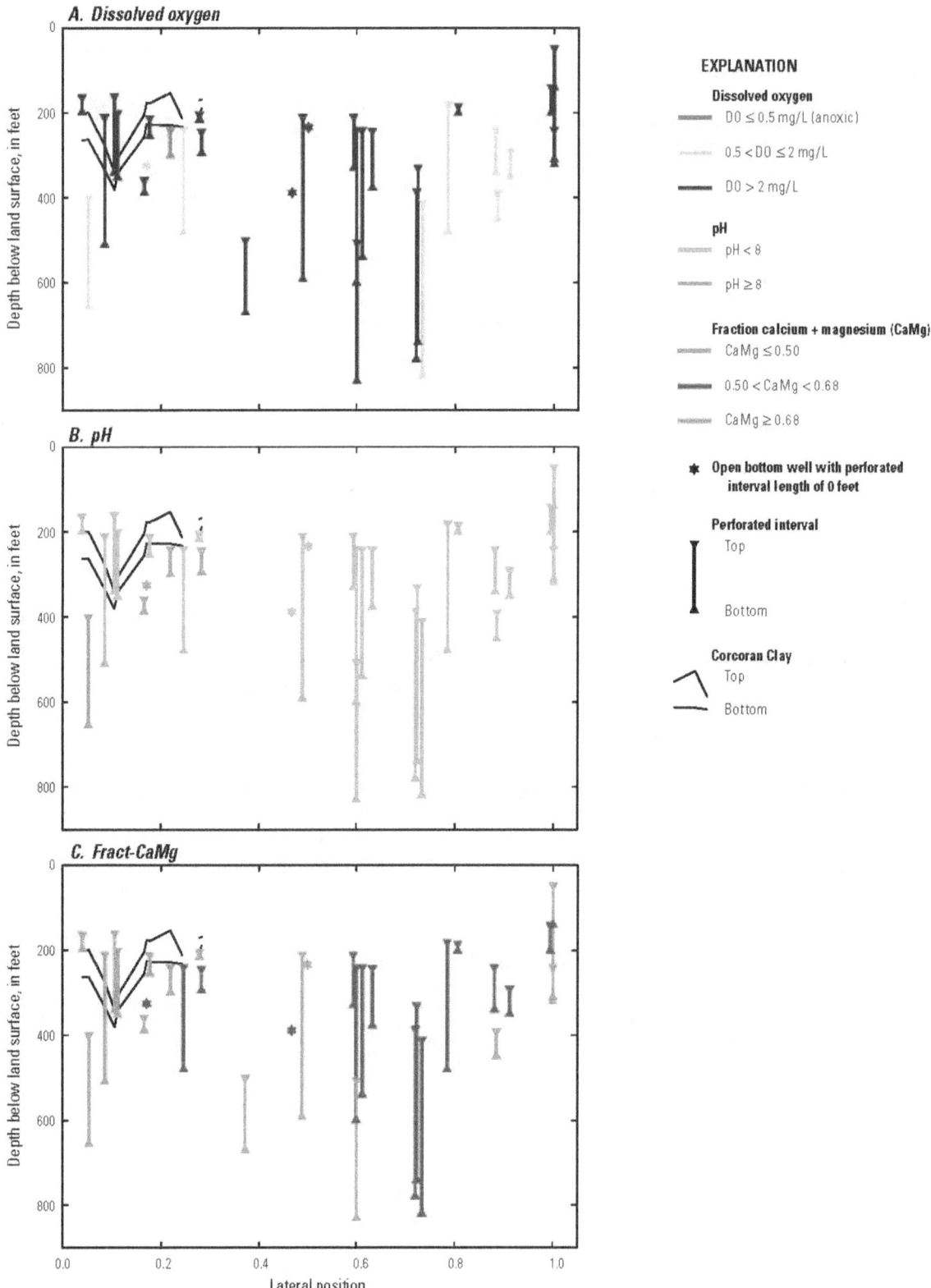

Figure 11. Relations of (*A*) dissolved oxygen, (*B*) pH, and (*C*) fraction of calcium+magnesium in total cations to lateral position and depth of perforated interval of wells, Madera-Chowchilla study unit, 2008, California GAMA Priority Basin Project.

Correlations Between Explanatory Variables

Apparent correlations between potential explanatory factors and water-quality constituents could result from correlations among potential explanatory factors; therefore, identification of statistically significant correlations between potential explanatory factors is important. The potential explanatory factors examined for this study unit are not extensively correlated with one another with the exception of the correlations between depth, groundwater age, position relative to the Corcoran Clay, and geochemical indicators (tables 6A,B).

There were relatively few significant correlations between land use and other potential explanatory factors. Agricultural land use had a significant negative correlation and natural land use a significant positive correlation with lateral position (table 6B). Natural land use is dominant in two areas of the study unit: along the edge of the Sierra Nevada foothills (eastern margin) and near the center of the western margin, close to the San Joaquin River. However, the natural land-use area near the River has little population and very few wells compared to the area along the foothills (fig. 6). Thus, for areas within the 500-meter (m) buffers around the grid wells, natural land use occurs nearly entirely in the area along the foothills (lateral position > 0.8). The significant association between agricultural land use and wells perforated above/across the Corcoran Clay (table 6A) is a consequence of the Clay only being present in the western side of the study unit and the relation between agricultural land use and lateral position.

The significant positive correlation between urban land use and well depth (table 6B) reflects the greater depth of wells serving larger populations. The wells sampled by USGS-GAMA can be divided into four types: CDPH wells serving populations greater than 4,000 people (n = 8), CDPH wells serving populations less than 500 people (n = 13), irrigation wells (n = 7), and domestic wells (n = 7). The median depth of CDPH wells serving large populations (670 ft) was significantly greater than the median depths of CDPH wells serving small populations (325 ft; p = 0.002), irrigation wells (346 ft, p = 0.008), or domestic wells (317 ft; p = 0.014) (see appendix B). There were no significant differences between the depths of CDPH wells serving small populations, irrigation wells, and domestic wells. The primary aquifer system is defined by the intervals over which wells in the CDPH database are perforated; therefore, all of the wells sampled by USGS-GAMA are representative of the primary aquifer system. CDPH wells serving large populations are located in or near the Cities of Fresno, Madera, and Chowchilla, or are located at institutions with large resident populations, which all correspond to areas with the most urban land use and the greatest population densities in the study unit. In contrast, CDPH wells serving small populations mostly are at schools, parks, stores, and restaurants that are located away from the major cities.

There were many significant correlations between well depth, groundwater age, and geochemical indicators. Groundwater with pre-modern age distributions was significantly associated with deeper wells and with greater depths to the top of the perforated interval (table 6A, figs. 9A,B). Groundwater with pre-modern age distributions also was significantly associated with higher pH and with lower Fract-CaMg values and dissolved oxygen concentrations. As expected from the relation between well depth and groundwater age, deeper wells had significantly higher pH and lower Fract-CaMg values and dissolved oxygen concentrations.

The lack of correlation between lateral position and dissolved oxygen was unexpected. Studies in other areas of the eastern San Joaquin Valley have found strong gradients in dissolved oxygen concentration in groundwater, with anoxic conditions commonly found in the center of the Valley (low lateral position) (Davis and Hall, 1959; Bertoldi and others, 1991; Burow and others, 1998a,b; Bennett and others, 2010; Landon and others, 2010).

Table 6A.   Results of Wilcoxon rank-sum tests for differences in values of selected potential explanatory factors between samples classified into groups by position relative to the Corcoran Clay, groundwater age, or well depth, Madera-Chowchilla study unit, 2008, California GAMA Priority Basin Project.

[Relation of median values in pairs of sample groups tested shown for Wilcoxon rank-sum tests in which the two populations were determined to be significantly different (two-sided test) on the basis of p-values (not shown) less than threshold value (α) of 0.05; ns, test indicates no significant difference between the pair of sample groups. **Position relative to Corcoran Clay:** Abv/Ac, bottom of perforations are above Corcoran Clay or well is perforated across the Corcoran Clay; East, well located east of Corcoran Clay extent in subsurface; Blw, topmost perforation is below the Corcoran Clay. **Age classification:** Mod, modern; Mix, mixed; Pre, pre-modern. **Depth classification:** Shal, entire perforated interval ≤ 280 ft; Mixed, top of perforated interval ≤ 280 ft and bottom of perforated interval > 280 ft; Deep, entire perforated interval > 280 ft. See Appendix A for more information about factors listed. **Other abbreviations:** Ca, calcium; Mg, magnesium; Fract-CaMg, calcium plus magnesium in milliequivalents divided by sum of calcium, magnesium, sodium, and potassium in milliequivalents; ft, feet below land surface; sig. diff., significant differences; <, less than; >, greater than]

Potential explanatory factor	Position relative to Corcoran Clay			Age classification			Depth classification		
	Abv/Ac vs Blw (sig. diff.)	Abv/Ac vs East (sig. diff.)	Blw vs East (sig. diff.)	Mod vs Pre (sig. diff.)	Mod vs Mix (sig. diff.)	Mix vs Pre (sig. diff.)	Shal vs Deep (sig. diff.)	Shal vs Mix (sig. diff.)	Deep vs Mix (sig. diff.)
Grid wells									
Agricultural land use (percent)	ns	Abv/Ac>East	ns	ns	ns	ns	ns	ns	ns
Natural land use (percent)	ns	East>AbvAc	ns	ns	ns	ns	ns	ns	ns
Urban land use (percent)	ns	ns	ns	ns	ns	Pre>Mix	ns	ns	ns
Orchard/vineyard land use (percent)	ns	ns	ns	ns	ns	ns	ns	ns	ns
Lateral position	ns	East>Abv/Ac	East>Blw	ns	ns	ns	ns	ns	ns
Grid and understanding wells									
Well depth	ns	ns	ns	ns	ns	Pre>Mix	Deep>Shal	Mix>Shal	Deep>Mix
Depth to top of perforation	Blw>Abv/Ac	East>Abv/Ac	ns	Pre>Mod	ns	ns	Deep>Shal	ns	Deep>Mix
pH	Blw>Abv/Ac	ns	ns	Pre>Mod	ns	ns	Deep>Shal	ns	ns
DO	Abv/Ac>Blw	Abv/Ac>East	ns	Mod>Pre	Mod>Mix	ns	Shal>Deep	Shal>Mix	ns
Fract-CaMg	Abv/Ac>Blw	Abv/Ac>East	ns	Mod>Pre	ns	ns	Shal>Deep	Shal>Mix	Mix>Deep

Table 6B. Results of Spearman's tests of correlations between selected potential explanatory factors, Madera-Chowchilla study unit, 2008, California GAMA Priority Basin Project.

[**Abbreviations:** ρ (rho), Spearman's correlation statistic; ρ values are shown for tests in which the variables were determined to be significantly correlated on the basis of p values (significance level of the Spearman's test) less than threshold value (α) of 0.05 (not shown); ns, Spearman's test indicates no significant correlation between factors; black text, significant positive correlation; red text, significant negative correlation]

ρ	Percent natural land use	Percent urban land use	Lateral position	Well depth	Depth to top of perforation	pH	Dissolved oxygen concentration	Fract-CaMg
Grid wells								
Percent agricultural land use	−0.82	−0.53	−0.66	ns	ns	ns	ns	ns
Percent natural land use		ns	0.72	ns	ns	ns	ns	ns
Percent urban land use			ns	0.40	ns	ns	ns	ns
Lateral position				ns	ns	ns	ns	ns
Grid and understanding wells								
Well depth					0.63	ns	ns	−0.43
Depth to top of perforation						0.53	−0.36	−0.61
pH							−0.44	−0.71
Dissolved oxygen concentration								0.60

Status and Understanding of Water Quality

The *status assessment* was designed to identify the constituents or classes of constituents most likely to be water-quality concerns by virtue of their high concentrations or their prevalence. The assessment applies only to constituents having regulatory or non-regulatory health-based or aesthetic/technical based benchmarks established by the USEPA or the CDPH (as of 2008). The spatially distributed, randomized approach to well selection and data analysis yields a view of groundwater quality in which all areas of the primary aquifer system are weighted equally; regions with a high density of groundwater use or with high density of potential contaminants were not preferentially sampled (Belitz and others, 2010).

The *understanding assessment* was designed to help answer the question of why specific constituents are, or are not, observed in groundwater in the area, and may improve our understanding of how human activities and natural processes affect groundwater quality in the study unit. The assessment addresses a subset of the constituents discussed in the *status assessment*, and is based on statistical correlations between water quality and a finite set of potential explanatory factors. The assessment is not designed to identify specific sources of constituents to specific wells.

The following discussion of the *status and understanding* assessment results is divided into two parts—inorganic constituents and organic constituents—and each part has a tiered structure. Each part begins with a survey of how many constituents were detected at any concentration in USGS-GAMA samples compared to the number analyzed, and a graphical summary of the RCs of constituents detected in the grid wells. Aquifer-scale proportions are presented for the subset of constituents that met criteria for additional evaluation based on RC, or for organic constituents, prevalence. *Understanding assessment* results are presented for the subset of *status assessment* constituents that had statistically significant correlations to potential explanatory factors. For constituents that have *understanding assessment* results, those results are presented immediately following the status assessment results for that constituent.

Inorganic Constituents

Inorganic constituents typically occur naturally in groundwater, although their concentrations may be influenced by human activities as well as by natural factors. Forty-four of the 46 inorganic constituents analyzed for by USGS-GAMA were detected (table 7A). Of these 44 constituents, 23 had regulatory or non-regulatory health-based benchmarks, 8 had non-regulatory aesthetic/technical-based benchmarks, and 13 had no established benchmarks. Of the 31 inorganic constituents with benchmarks, 11 were identified for additional evaluation in the *status assessment* because they were detected at moderate or high RCs in the grid wells: nitrate, arsenic, barium, uranium, vanadium, gross alpha particle activity, uranium activity (sum of uranium-234, uranium-235, and uranium-238), manganese, chloride, total dissolved solids (TDS), and specific conductance (fig. 12). The majority of these 11 constituents were detected at moderate or high RCs in more than 15% of the grid wells (fig. 13).

Table 7A.   Number of inorganic constituents analyzed and detected, by benchmark type and constituent type, Madera-Chowchilla study unit, 2008, California GAMA Priority Basin Project.

[Health-based benchmarks (HBB) include USEPA maximum contaminant level, CDPH maximum contaminant level, USEPA lifetime health advisory levels, action levels, and CDPH notification levels. CDPH secondary maximum contaminant level benchmarks (SMCL) are non-regulatory aesthetic benchmarks. **Abbreviations:** USEPA, U.S. Environmental Protection Agency; CDPH, California Department of Public Health]

Benchmark type	Number analyzed	Number detected
Nutrients		
HBB	3	3
SMCL	0	0
No benchmark	2	2
Total:	5	5
Major and minor ions		
HBB	1	1
SMCL	4	4
No benchmark	8	8
Total:	13	13
Trace elements		
HBB	18	16
SMCL	4	4
No benchmark	3	3
Total:	25	23
Radioactive constituents		
HBB	3	3
SMCL	0	0
No benchmark	0	0
Total:	3	3
Sum of inorganic constituents		
HBB	25	23
SMCL	8	8
No benchmark	13	13
Total:	46	44

Table 7B.   Number of organic and special-interest constituents analyzed and detected, by health-based benchmark type and constituent type, Madera-Chowchilla study unit, 2008, California GAMA Priority Basin Project.

[Regulatory health-based benchmarks (HBB) include USEPA maximum contaminant level and CDPH maximum contaminant level. Non-regulatory health-based benchmarks include USEPA lifetime health advisory levels, risk-specific dose level at 10^{-5}, and CDPH notification level. **Abbreviations:** USEPA, U.S. Environmental Protection Agency; CDPH, California Department of Public Health; VOC, volatile organic compound]

Benchmark type	Number analyzed	Number detected
Organic constituents		
Pesticides and pesticide degradates		
Regulatory - HBB	14	3
Non-regulatory - HBB	29	4
No benchmark	91	5
Total:	134	12
Fumigants		
Regulatory - HBB	4	3
Non-regulatory - HBB	4	1
No benchmark	2	0
Total:	10	4
Other VOCs		
Regulatory - HBB	29	8
Non-regulatory - HBB	21	0
No benchmark	25	0
Total:	75	8
Constituents of special interest		
Regulatory - HBB	1	1
Non-regulatory - HBB	1	1
No benchmark	0	0
Total:	2	2
Sum of organic and special-interest constituents		
Regulatory - HBB	48	15
Non-regulatory - HBB	55	6
No benchmark	118	5
Total:	221	26

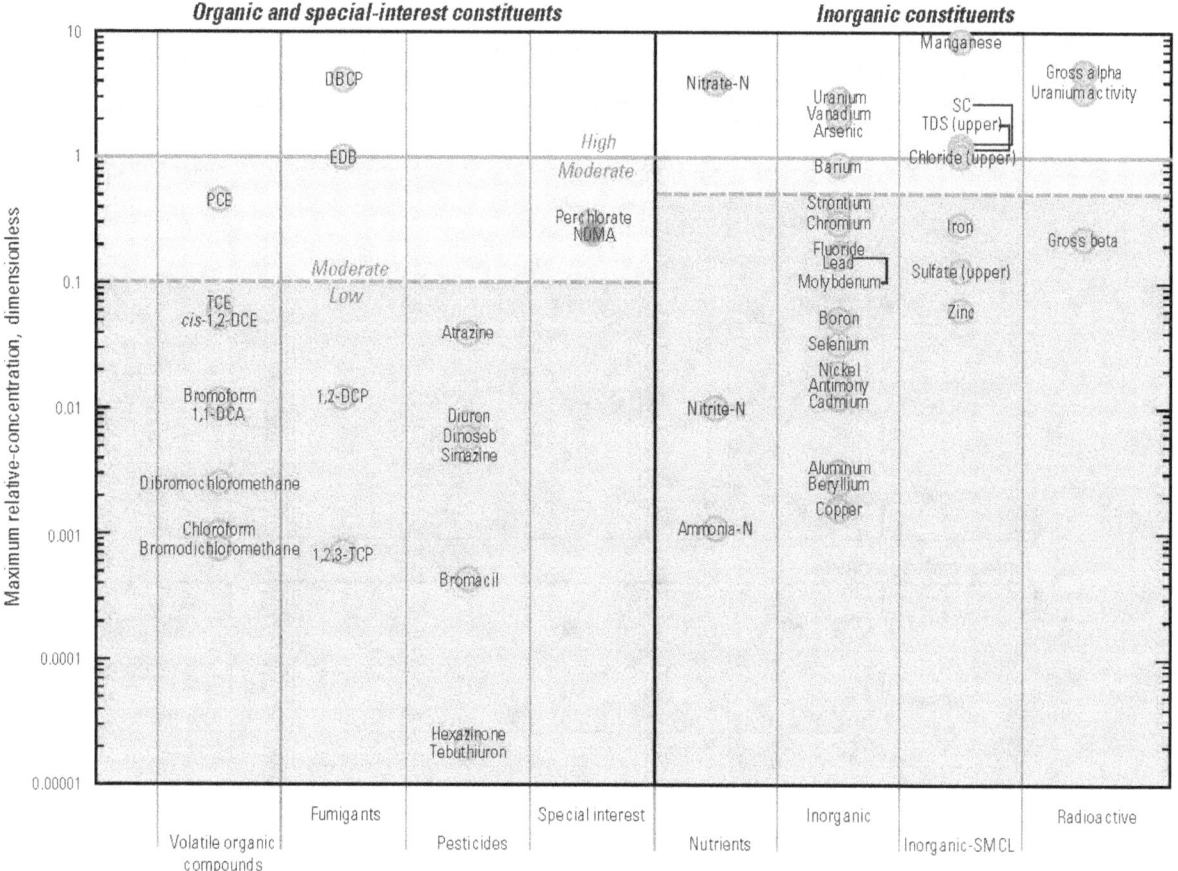

EXPLANATION

PCE   **Constituents with analyses in >20 grid wells and wells are spatially representative—**
        Name and center of symbol is location of data unless indicated by following location line:

NDMA   **Constituent with analyses in <20 grid wells and wells are not spatially representative—**
        Name and center of symbol is location of data unless indicated by following location line:

Abbreviations

cis-1,2-DCE, cis-1,2-dichloroethene; DBCP, 1,2-dibromo-3-chloropropane; EDB, 1,2-dibromoethane; 1,1-DCA, 1,1-dichloroethane; 1,2-DCP, 1,2-dichloropropane; N, as nitrogen; NDMA, N-nitrosodimethylamine; PCE, tetrachloroethene; TCE, trichloroethene; 1,2,3-TCP, 1,2,3-trichloropropane; SMCL, secondary maximum contaminant level; SC, specific conductance; TDS (upper), total dissolved solids; upper, upper water-quality benchmark where multiple benchmarks exist.

Figure 12.   Maximum relative-concentrations of constituents detected in grid wells, by type of constituent, in the Madera-Chowchilla study unit, 2008, California GAMA Priority Basin Project.

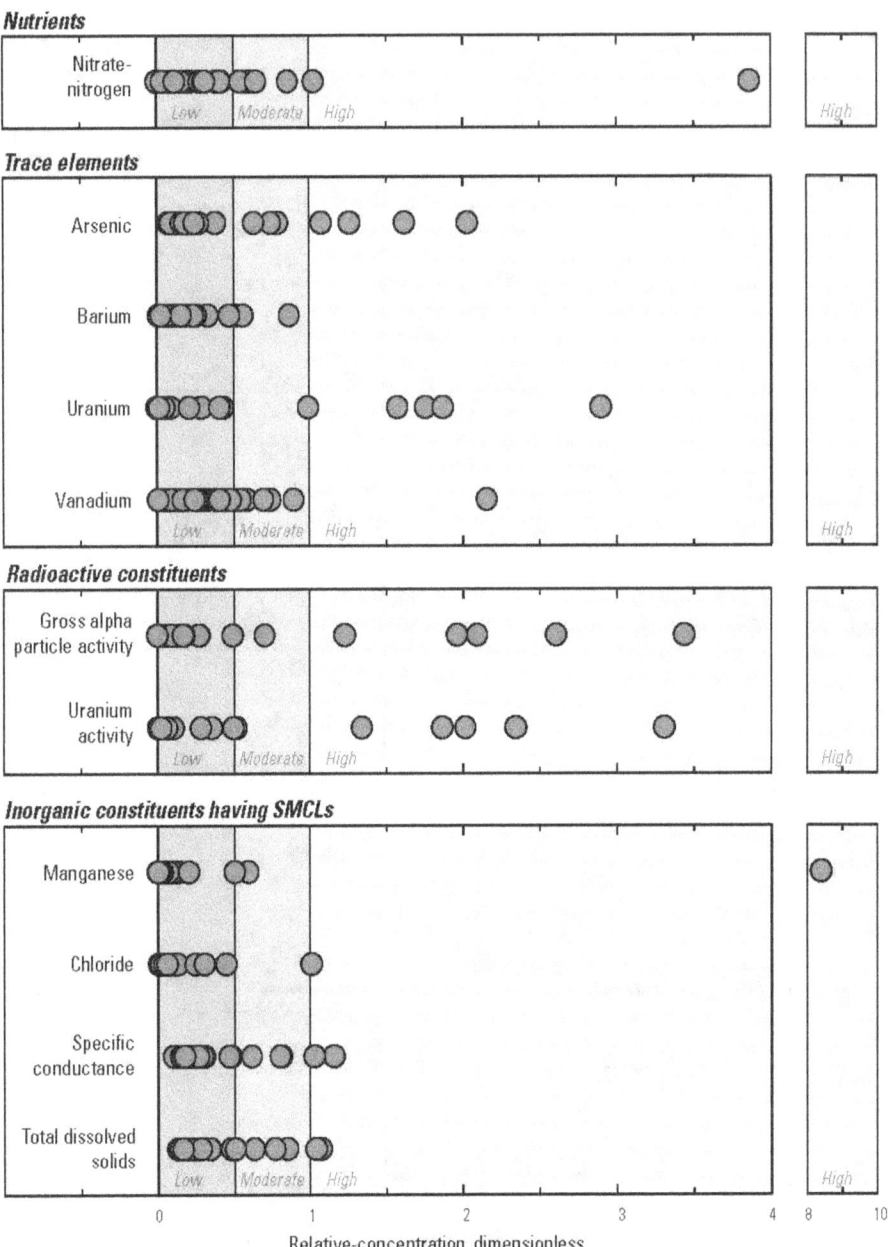

Figure 13.   Relative-concentrations of selected nutrients, trace elements, radioactive constituents, and inorganic constituents with secondary maximum contaminant levels detected in grid wells, Madera-Chowchilla study unit, 2008, California GAMA Priority Basin Project.

Although uranium was considered a radioactive constituent for the *status assessment*, uranium is shown as both a radioactive constituent and a trace element in figures 12 and 13 to demonstrate that similar results were obtained for comparison to the MCL-CA of 20 picocuries per liter (pCi/L) and the MCL-US of 30 micrograms per liter (µg/L). Similarly, TDS and specific conductance are both shown on figures 12 and 13 to demonstrate that similar results were obtained for these two measures of salinity.

Three additional constituents were selected for additional evaluation in the *status assessment* because they had moderate or high RCs in the datasets used for the spatially weighted calculations of aquifer-scale proportions: strontium, lead, and iron (table 8). Strontium was included because it was reported at a high RC in one of the understanding wells. Lead and iron were included because they were reported at moderate or high RCs in the CDPH database during the 3-year interval prior to USGS sampling.

Aquifer-scale proportions for the constituents selected for additional evaluation in the *status assessment* are summarized in table 8. Spatially weighted high aquifer-scale proportions fell within the 90% confidence intervals for their respective grid-based aquifer high proportions for all constituents except iron, providing evidence that the grid-based approach yields statistically equivalent results to the spatially weighted approach. Aquifer-scale proportions for classes of inorganic constituents are summarized in table 9A.

For any inorganic constituent having health-based benchmarks (nutrients, trace elements, and radioactive constituents), 37% of the primary aquifer system had high RCs of at least one constituent, 30% had moderate values, and 33% had low values (table 9A). High RCs of nutrients, trace elements, and radioactive constituents all contributed to the high aquifer-scale proportions for inorganic constituents. For any inorganic constituent having non-health-based benchmarks (SMCL constituents), 6.7% of the primary aquifer system had high RCs of at least one constituent (table 9A). High RCs of TDS accounted for most of the high aquifer-scale proportion.

Nutrients

Nitrate was the only nutrient detected at moderate or high RCs in the study unit (table 8). Nitrate was detected at high RCs in 6.7% of the primary aquifer system and at moderate RCs in 20% (table 9A; fig. 13). Wells with high and moderate RCs of nitrate were distributed across the study unit (fig. 14).

Factors Affecting Nitrate

Groundwater age, depth, and geochemical conditions were the most significant explanatory factors related to nitrate concentrations. Nitrate concentrations were significantly higher in wells with modern groundwater, compared to pre-modern groundwater (table 10A), and had significant negative correlations with well depths and depths to top of perforations (table 10B; fig. 15A). All of the wells with high or

moderate RCs of nitrate had depth to top of perforations less than or equal to 240 ft below land surface (fig. 15A). Nitrate concentrations were significantly higher in wells perforated above/across the Corcoran Clay compared to wells perforated below the Clay or wells east of the Clay (table 10A), likely because all of the wells classified as above/across the Corcoran had depths to top of perforations less than 240 ft. Nitrate concentrations were not significantly associated with lateral position (table 10B).

Nitrate concentration had a significant positive correlation with dissolved oxygen concentration and TDS (figs. 16A,B), and a negative correlation with pH (table 10B). These correlations likely reflect the correlations between geochemical conditions and well depth—dissolved oxygen had a significant negative correlation and pH a significant positive correlation with depth to top of perforations (table 6B). Denitrification of nitrate to nitrogen gas and intermediate products has been identified in other areas of the eastern San Joaquin Valley (McMahon and others, 2008; Landon and others, 2010). In these studies, however, denitrification was associated with reducing conditions— conditions that are rare in the Madera-Chowchilla study unit (fig. 11A). Thus, the low RCs of nitrate found in deep wells with mixed and pre-modern, oxic groundwater (fig. 16A) and found in groundwater with low RCs of TDS (fig. 16B) likely reflect the initial nitrate concentrations in the groundwater recharge. Modern recharge appears to have a much wider range of nitrate concentrations.

The associations between higher nitrate concentrations, shallower wells, well-oxygenated conditions, higher TDS, and recent (modern) recharge are similar to those found in other studies of the eastern San Joaquin Valley (Burow and others, 1998a,b; Dubrovsky and others, 1998). These other studies also reported significant positive correlations between nitrate concentrations and percentage of agricultural land use in the vicinity of wells. Evaluation of historical datasets for nitrate concentrations in shallow groundwater (<200 ft) in the eastern San Joaquin Valley indicated that nitrate concentrations increased significantly from the 1950s to the 1980s, which approximately correlates with the increase in the amount of nitrate fertilizer applied in the eastern San Joaquin Valley (Dubrovsky and others, 1998; Burow and others, 2007).

Nitrate concentrations were not correlated with percentage of agricultural land use and were inversely correlated with percentage of urban land use in the Madera-Chowchilla study unit (table 10B). The inverse correlation to urban land use may be explained by the association between more urbanized areas with greater population densities and deeper wells (see appendix B). The lack of correlation between nitrate and agricultural land use may be explained by the relatively large number of deep wells with pre-modern groundwater sampled for this study (fig. 15A). Previous USGS studies in the eastern San Joaquin Valley have focused primarily on shallower parts of the aquifer system and therefore sampled a greater proportion of wells receiving modern recharge.

Table 8. Aquifer-scale proportions using grid-based and spatially weighted methods for those constituents that met criteria for additional evaluation in the status assessment, Madera-Chowchilla study unit, 2008, California GAMA Priority Basin Project.

[Grid-based aquifer-scale proportions are based on samples collected by the USGS from 30 grid wells during April–May 2008. Spatially weighted aquifer-scale proportions and raw detection frequencies are based on samples collected by the USGS from the 30 grid wells and 5 understanding wells during April–May 2008 and data reported in the California Department of Public Health database during February 12, 2005–February 12, 2008. **Relative-concentration categories: high,** concentrations greater than water-quality benchmark; **moderate,** concentrations greater than or equal to 0.1 of benchmark but less than benchmark for organic constituents (threshold for inorganic constituents is 0.5 of benchmark); **low,** concentrations less than 0.1 of benchmark for organic constituents (threshold for inorganic constituents is 0.5 of benchmark). Benchmark types and values listed in table 3. **Other abbreviations:** USGS, U.S. Geological Survey; VOC, volatile organic compound; DBCP, 1,2-dibromo-3-chloropropane; EDB, 1,2-dibromoethane; PCE, tetrachloroethene; 123-TCP, 1,2,3-trichloropropane; TDS, total dissolved solids; THM, trihalomethane; SMCL, secondary maximum contaminant level; NA, not applicable]

Constituent	Raw detection frequency[1]			Spatially weighted aquifer-scale proportion[1]			Grid-based aquifer-scale proportion			90 percent confidence interval for grid-based high proportion[2]	
	Number of wells	Percent moderate	Percent high	Number of cells	Moderate proportion (percent)	High proportion (percent)	Number of cells	Moderate proportion (percent)	High proportion (percent)	Lower limit (percent)	Upper limit (percent)
Nutrients											
Nitrate	131	15	5.3	30	24	9.7	30	20	6.7	1.9	17
Trace elements				Inorganic constituents with health-based benchmarks							
Arsenic	84	4.8	9.5	30	6.0	9.8	30	10	13	5.7	26
Barium	83	4.8	1.2	30	9.0	1.1	30	10	0	0	4.4
Lead	82	1.2	0	30	0.5	0	30	0	0	0	4.4
Strontium	35	0	2.9	30	0	1.7	30	0	0	0	4.4
Uranium	62	6.4	11	30	7.2	17	30	3.3	13	5.7	26
Vanadium	42	17	2.4	30	20	1.1	30	20	3.3	0.6	12
Radioactive constituents											
Gross alpha particle activity[3]	97	3.1	10	30	4.5	23	30	3.3	20	10	34
Uranium activity	49	6.1	16	28	4.8	20	30	3.3	17	7.9	30
Salinity indicators				Inorganic constituents with aesthetic or technical (SMCL) benchmarks							
Chloride	91	0	2.2	30	0	4.4	30	0	3.3	0.6	12
TDS	91	8	3.3	30	17	7.8	30	13	6.7	1.9	17
Specific conductance	102	13	2.9	30	12	7.8	30	10	6.7	1.9	17
Trace elements											
Iron	91	1.1	11	30	1.1	7.0	30	0	0	0	4.4
Manganese	91	5.5	5.5	30	4.6	5.9	30	3.3	3.3	0.6	12
Herbicides				Organic and special-interest constituents							
Atrazine[4]	94	0	0	30	0	0	30	0	0	0	4.4
Diuron[4]	36	0	0	29	0	0	29	0	0	0	4.5
Simazine[4]	94	0	0	30	0	0	30	0	0	0	4.4

Table 8. Aquifer-scale proportions using grid-based and spatially weighted methods for those constituents that met criteria for additional evaluation in the status assessment, Madera-Chowchilla study unit, 2008, California GAMA Priority Basin Project.—Continued

[Grid-based aquifer-scale proportions are based on samples collected by the USGS from 30 grid wells during April–May 2008. Spatially weighted aquifer-scale proportions and raw detection frequencies are based on samples collected by the USGS from the 30 grid wells and 5 understanding wells during April–May 2008 and data reported in the California Department of Public Health database during February 12, 2005–February 12, 2008. **Relative-concentration categories: high,** concentrations greater than water-quality benchmark; **moderate,** concentrations greater than or equal to 0.1 of benchmark but less than benchmark for organic constituents (threshold for inorganic constituents is 0.5 of benchmark); **low,** concentrations less than 0.1 of benchmark for organic constituents (threshold for inorganic constituents is 0.5 of benchmark). Benchmark types and values listed in table 3. **Other abbreviations:** USGS, U.S. Geological Survey; VOC, volatile organic compound; DBCP, 1,2-dibromo-3-chloropropane; EDB, 1,2-dibromoethane; PCE, tetrachloroethene; 123-TCP, 1,2,3-trichloropropane; TDS, total dissolved solids; THM, trihalomethane; SMCL, secondary maximum contaminant level; NA, not applicable]

Constituent	Raw detection frequency[1]			Spatially weighted aquifer-scale proportion[1]			Grid-based aquifer-scale proportion			90 percent confidence interval for grid-based high proportion[2]	
	Number of wells	Percent moderate	Percent high	Number of cells	Moderate proportion (percent)	High proportion (percent)	Number of cells	Moderate proportion (percent)	High proportion (percent)	Lower limit (percent)	Upper limit (percent)
				Organic and special-interest constituents—Continued							
Fumigants											
DBCP[4]	90	1.1	3.3	30	0	3.9	30	0	10	3.7	22
EDB[4]	90	1.1	0	30	1.1	0	30	3.3	0	0	4.4
1,2,3-TCP[4]	81	0	0	30	0	0	30	0	0	0	4.4
Other VOCs (THMs and solvents)											
Chloroform[4]	81	0	0	30	0	0	30	0	0	0	4.4
PCE[4]	79	2.5	0	30	0.4	0	30	3.3	0	0	4.4
Constituents of special interest											
NDMA[5]	10	10	0	5	NA	NA	5	NA	NA	NA	NA
Perchlorate[4]	72	8.3	0	30	16	0	30	20	0	0	4.4

[1] Based on most recent CDPH analysis during February 12, 2005–February 12, 2008, combined with grid-based data.

[2] Based on the Jeffreys interval for the binomial distribution (Brown and others, 2001).

[3] Gross alpha particle activities were not adjusted for uranium activity. The MCL-US for gross alpha particle activity applies to adjusted gross alpha particle activity, which is equal to measured gross alpha particle activity minus uranium activity.

[4] Organic constituent or constituent of special interest detected in at least 10 percent of grid wells.

[5] Aquifer-scale proportions were not calculated for NDMA because only 5 of 30 grid cells had data for NDMA.

Table 9A.   Aquifer-scale proportions for inorganic constituent classes, Madera-Chowchilla study unit, 2008, California GAMA Priority Basin Project.

[**Relative-concentration categories:** high, concentration of at least one constituent in group greater than water-quality benchmark; moderate, concentration of at least one constituent in group greater than 0.5 of benchmark and no constituents in group with concentration greater than benchmark; low, concentrations of all constituents in group less than or equal to 0.5 of benchmark. **Abbreviations:** SMCL, secondary maximum contaminant level; TDS, total dissolved solids; SC, specific conductance]

Constituent class	Aquifer-scale proportions		
	Low relative-concentrations (percent)	Moderate relative-concentrations (percent)	High relative-concentrations (percent)
Inorganic constituents with health-based benchmarks			
Nutrients	73	20	6.7
Trace elements [1]	57	30	13
Uranium and radioactive constituents [1]	77	3.3	20
Any inorganic constituent with a health-based benchmark	33	30	37
Inorganic constituents with aesthetic or technical (SMCL) benchmarks			
Salinity indicators (TDS, SC, chloride)	80	13	6.7
Manganese	93	3.3	3.3
Any inorganic constituent with an SMCL benchmark	77	17	6.7

[1] Uranium is not included in the trace element class.

[2] Aquifer-scale proportions for the class uranium and radioactive constituents were calculated using unadjusted gross alpha particle activity.

Table 9B.   Aquifer-scale proportions for organic constituent classes, Madera-Chowchilla study unit, 2008, California GAMA Priority Basin Project.

[**Relative-concentration categories:** high, concentration of at least one constituent in group greater than water-quality benchmark; moderate, concentration of at least one constituent in group greater than 0.1 of benchmark and no constituents in group with concentration greater than benchmark; low, concentrations of all constituents in group less than or equal to 0.1 of benchmark. Abbreviation: THM, trihalomethane]

Constituent class	Aquifer-scale proportions			
	Not detected (percent)	Low relative-concentrations (percent)	Moderate relative-concentrations (percent)	High relative-concentrations (percent)
Organic constituents with regulatory and non-regulatory health-based benchmarks				
Herbicides	77	23	0	0
Fumigants	66	24	[1]0.2	10
Solvents	90	6.7	3.3	0
THMs	83	17	0	0
Any organic constituent	47	40	3.3	10

[1] Spatially weighted

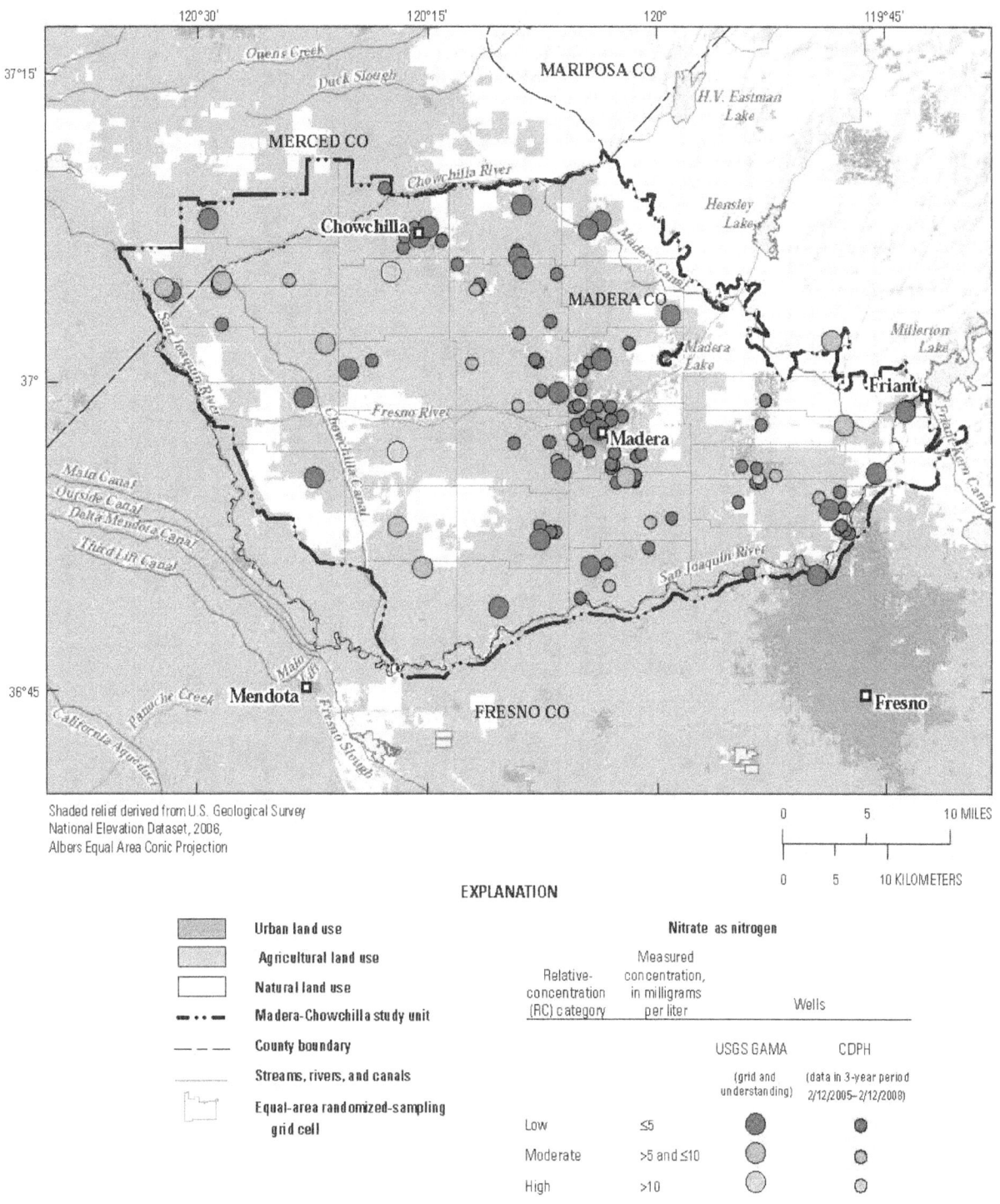

Figure 14. Concentrations of nitrate, as nitrogen, in USGS-GAMA wells and the most recent analysis during February 12, 2005–February 12, 2008, for CDPH wells, Madera-Chowchilla study unit, California GAMA Priority Basin Project.

Table 10A.   Results of Wilcoxon rank-sum tests for differences in values of water-quality constituents between samples classified into groups by position relative to the Corcoran Clay, groundwater age, or well depth, Madera-Chowchilla study unit, 2008, California GAMA Priority Basin Project.

[Relation of median values in pairs of sample groups tested shown for Wilcoxon rank-sum tests in which the two populations were determined to be significantly different (two-sided test) on the basis of p-values (not shown) less than threshold value (α) of 0.05; ns, test indicates no significant difference between the pair of sample groups. **Position relative to Corcoran Clay:** Abv/Ac, bottom of perforations above Corcoran Clay or well is perforated across the Corcoran Clay: East, well located east of Corcoran Clay extent in subsurface; Blw, topmost perforation is below the Corcoran Clay. **Age classification:** Mod, modern; Mix, mixed; Pre, pre-modern. **Depth classification:** Shal, entire perforated interval and well depth $\leq$ 280 ft, Mixed, top of perforations $\leq$ 280 ft, bottom of perforations and well depth > 280 ft, Deep, entire perforated interval and well depth > 280 ft. See Appendix A for more information about explanatory factor listed. **Other abbreviations:** TDS, total dissolved solids; THM, trihalomethane; ft, foot; sig. diff., significant differences; <, less than; >, greater than]

Water-quality constituents	Position relative to Corcoran Clay			Age classification			Depth classification		
	Abv/Ac vs Blw (sig. diff.)	Abv/Ac vs East (sig. diff.)	Blw vs East (sig. diff.)	Mod vs Pre (sig. diff.)	Mod vs Mix (sig. diff.)	Mix vs Pre (sig. diff.)	Shal vs Deep (sig. diff.)	Shal vs Mix (sig. diff.)	Deep vs Mix (sig. diff.)
Inorganic constituents with health-based benchmarks									
Nitrate	Abv/Ac>Blw	Abv/Ac>East	ns	Mod>Pre	ns	ns	Shal>Deep	Shal>Mix	ns
Arsenic	ns	ns	ns	ns	ns	ns	Deep>Shal	ns	Deep>Mix
Vanadium	ns	ns	ns	Mod>Pre	ns	ns	ns	ns	ns
Gross alpha particle activity	Abv/Ac>Blw	Abv/Ac>East	ns	Mod>Pre	Mod>Mix	ns	ns	ns	Mix>Deep
Uranium activity	Abv/Ac>Blw	Abv/Ac>East	ns	Mod>Pre	Mod>Mix	ns	Shal>Deep	ns	ns
Inorganic constituents with aesthetic or technical (SMCL) benchmarks									
Chloride	Abv/Ac>Blw	Abv/Ac>East	Blw>East	Mod>Pre	ns	ns	ns	ns	ns
TDS	Abv/Ac>Blw	Abv/Ac>East	ns	Mod>Pre	ns	ns	Shal>Deep	ns	Mix>Deep
Manganese	ns	ns	ns	Pre>Mod	ns	Pre>Mix	ns	ns	ns
Organic constituents and constituent of special interest									
Sum of herbicides	ns	ns	ns	Mod>Pre	Mod>Mix	ns	Shal>Deep	ns	Mix>Deep
Sum of fumigants	ns	ns	ns	ns	ns	ns	ns	ns	ns
Sum of THMs	ns	ns	ns	Mod>Pre	ns	ns	ns	ns	ns
Sum of solvents	ns	ns	ns	ns	Mod>Mix	ns	ns	ns	ns
Perchlorate	ns	ns	ns	ns	ns	ns	Shal>Deep	ns	ns

Table 10B. Results of Spearman's tests of correlations between selected potential explanatory factors and water-quality constituents, Madera-Chowchilla study unit, 2008, California GAMA Priority Basin Project.

[Abbreviations: ρ (rho), Spearman's correlation statistic; p values are shown for tests in which the variables were determined to be significantly correlated on the basis of p-values (significance level of the Spearman's test) less than threshold value (α) of 0.05 (not shown); ns, Spearman's test indicates no significant correlation between factors; black text, significant positive correlation; red text, significant negative correlation. FractCaMg, calcium plus magnesium divided by sum of calcium, magnesium, sodium, and potassium in milliequivalents]

ρ	Grid wells				Grid and understanding wells				
	Percent agricultural land use	Percent natural land use	Percent urban land use	Lateral position	Well depth	Depth to top of perforation	pH	Dissolved oxygen concentration	Fract-CaMg
Inorganic constituents with health-based benchmarks									
Nitrate	ns	ns	−0.41	ns	−0.49	−0.48	−0.56	0.73	0.66
Arsenic	ns	ns	ns	ns	ns	ns	0.44	−0.36	ns
Vanadium	ns	ns	ns	ns	ns	ns	ns	ns	ns
Gross alpha particle activity	ns	−0.40	ns	−0.46	ns	−0.60	ns	0.36	0.46
Uranium activity	0.38	ns	ns	ns	ns	−0.61	−0.45	0.47	0.63
Inorganic constituents with aesthetic or technical (SMCL) benchmarks									
Chloride	0.59	−0.63	ns	−0.69	ns	ns	ns	0.38	ns
Total dissolved solids	ns	ns	ns	ns	−0.43	−0.51	−0.43	0.42	0.62
Manganese	ns	ns	0.54	ns	ns	0.37	0.41	ns	ns
Organic constituents and constituent of special interest									
Sum of herbicides	ns	ns	ns	ns	ns	−0.46	−0.43	0.43	0.46
Sum of fumigants	ns	ns	ns	ns	ns	ns	ns	ns	ns
Sum of trihalomethanes	ns	ns	0.52	ns	ns	ns	ns	ns	ns
Sum of solvents	ns	ns	0.43	ns	ns	ns	ns	ns	ns
Perchlorate	ns	ns	ns	ns	−0.41	ns	−0.38	0.62	0.45

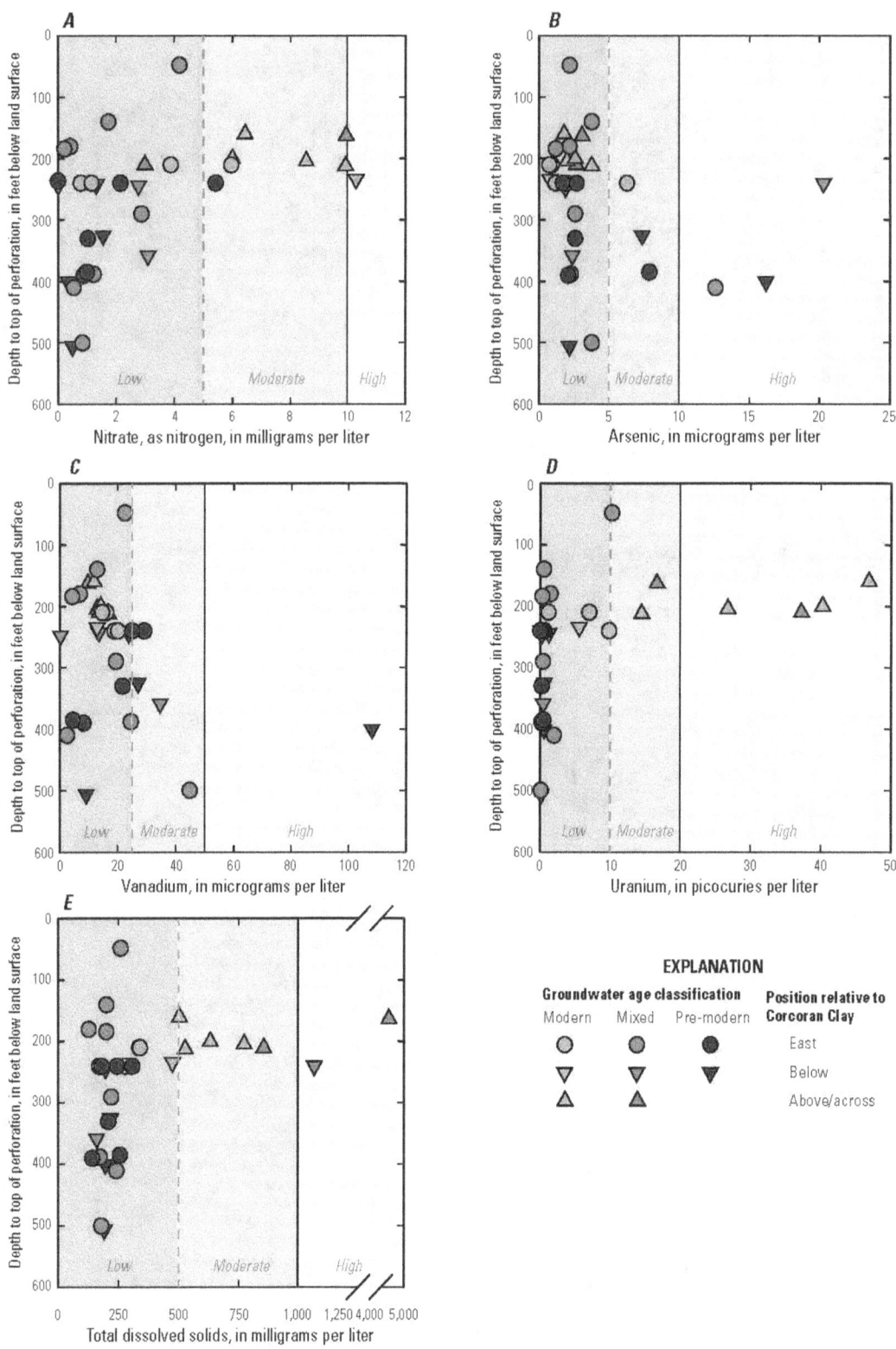

Figure 15.   Relation between concentrations of (*A*) nitrate, (*B*) arsenic, (*C*) vanadium, (*D*) uranium activity, and (*E*) total dissolved solids with depth to top of perforations in wells, Madera-Chowchilla study unit, 2008, California GAMA Priority Basin Project.

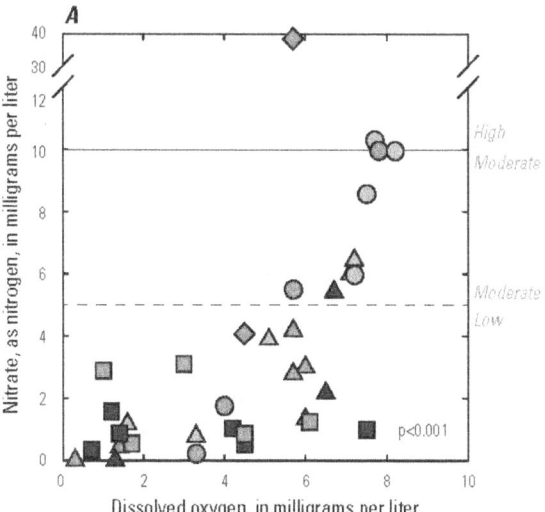

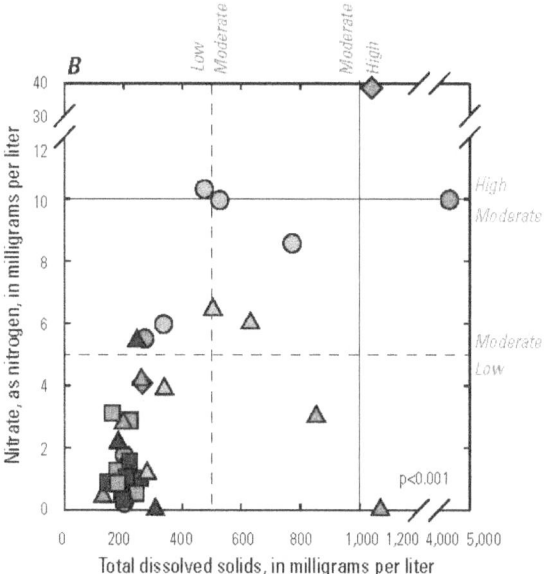

EXPLANATION

Groundwater age classification

Modern   Mixed   Pre-modern

Well depth classification

◎ (Modern) ◉ (Mixed)   Shallow (Entire perforated interval ≤ 280 feet)

△ (Modern) △ (Mixed) ▲ (Pre-modern)   Mixed (Top of perforated interval ≤ 280 feet; Bottom of perforated interval > 280 feet)

▨ (Mixed) ■ (Pre-modern)   Deep (Entire perforated interval > 280 feet)

◆   Unknown

Figure 16. Relations of (*A*) nitrate, as nitrogen concentration, and dissolved oxygen concentration, and (*B*) nitrate, as nitrogen concentration, and total dissolved solids, Madera-Chowchilla study unit, 2008, California GAMA Priority Basin Project.

Trace Elements

The constituent class trace elements includes a variety of metallic and non-metallic constituents that typically are present in groundwater at concentrations less than 1 mg/L. Trace elements with health-based benchmarks had high RCs in 13% of the primary aquifer system, moderate RCs in 30%, and low RCs in 57% (table 9A). Arsenic accounted for most of the high and moderate RCs of trace elements with health-based benchmarks (table 8; fig. 13).

Arsenic was detected at high RCs in 13% of the primary aquifer system and at moderate RCs in 10% (table 8, fig. 13). High and moderate RCs of arsenic primarily occurred in the northern part of the study unit (fig. 17). Vanadium was detected at high RCs in 3.3% of the primary aquifer system and at moderate RCs in 20% (table 8, fig. 13). The well with a high RC of vanadium was in the northwestern corner of the study unit, and wells with moderate RCs of vanadium were distributed throughout the study unit (fig. 18). Strontium was detected at high RCs in 1.7% of the primary aquifer system (spatially weighted), and at moderate RCs in 0%. The USGS-GAMA understanding well with a high RC of strontium was located in the northwestern corner of the study unit in the same sample as the high RC value of barium (fig. 19). Barium was detected at high RCs in 1.1% of the primary aquifer system (spatially weighted) and at moderate RCs in 10%. The high and moderate RCs of barium were predominantly located on the western side of the study unit (fig. 19). Uranium is discussed with radioactive constituents.

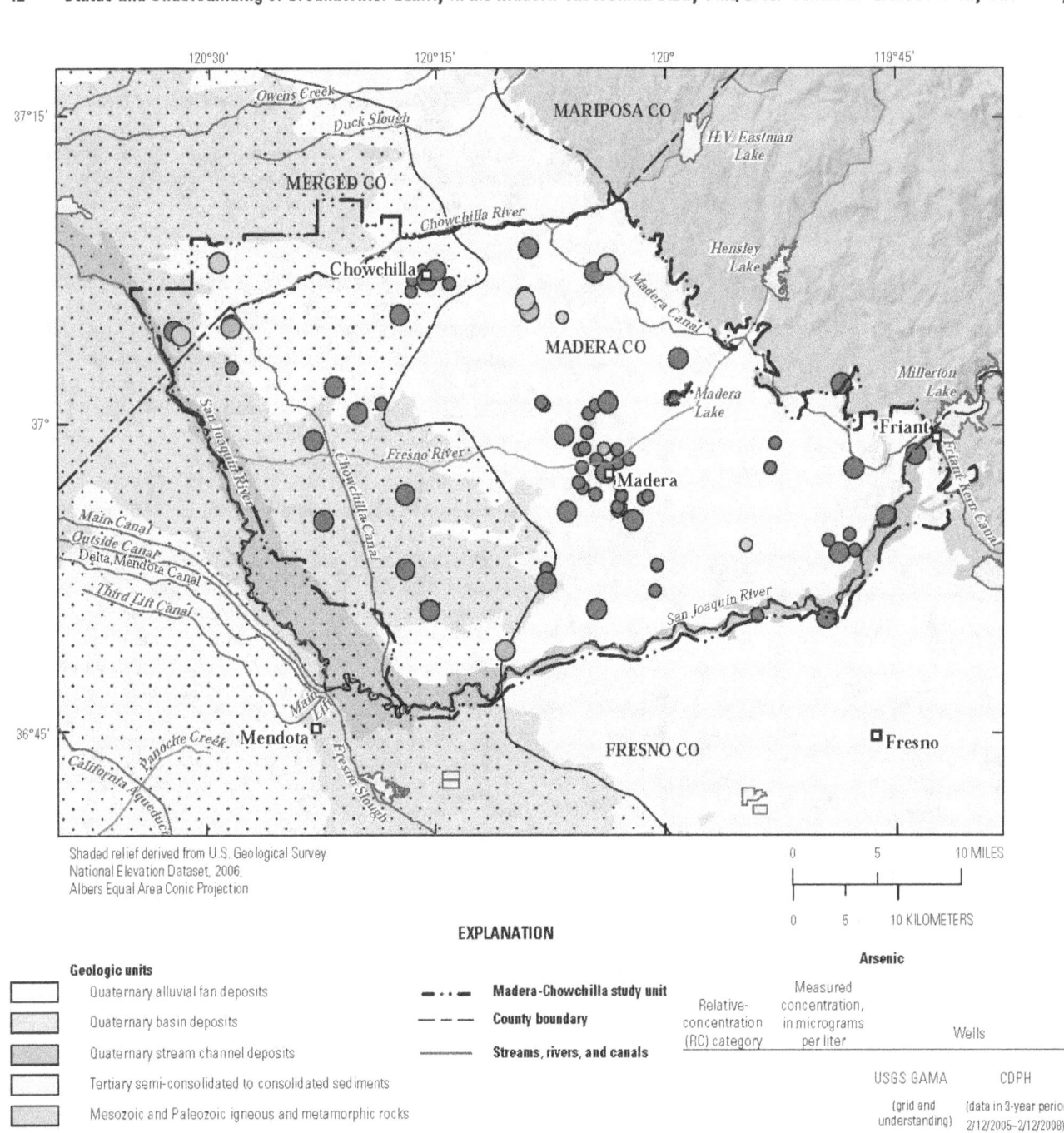

Figure 17.   Concentrations of arsenic in USGS-GAMA wells and the most recent analysis during February 12, 2005–February 12, 2008, for CDPH wells, Madera-Chowchilla study unit, California GAMA Priority Basin Project.

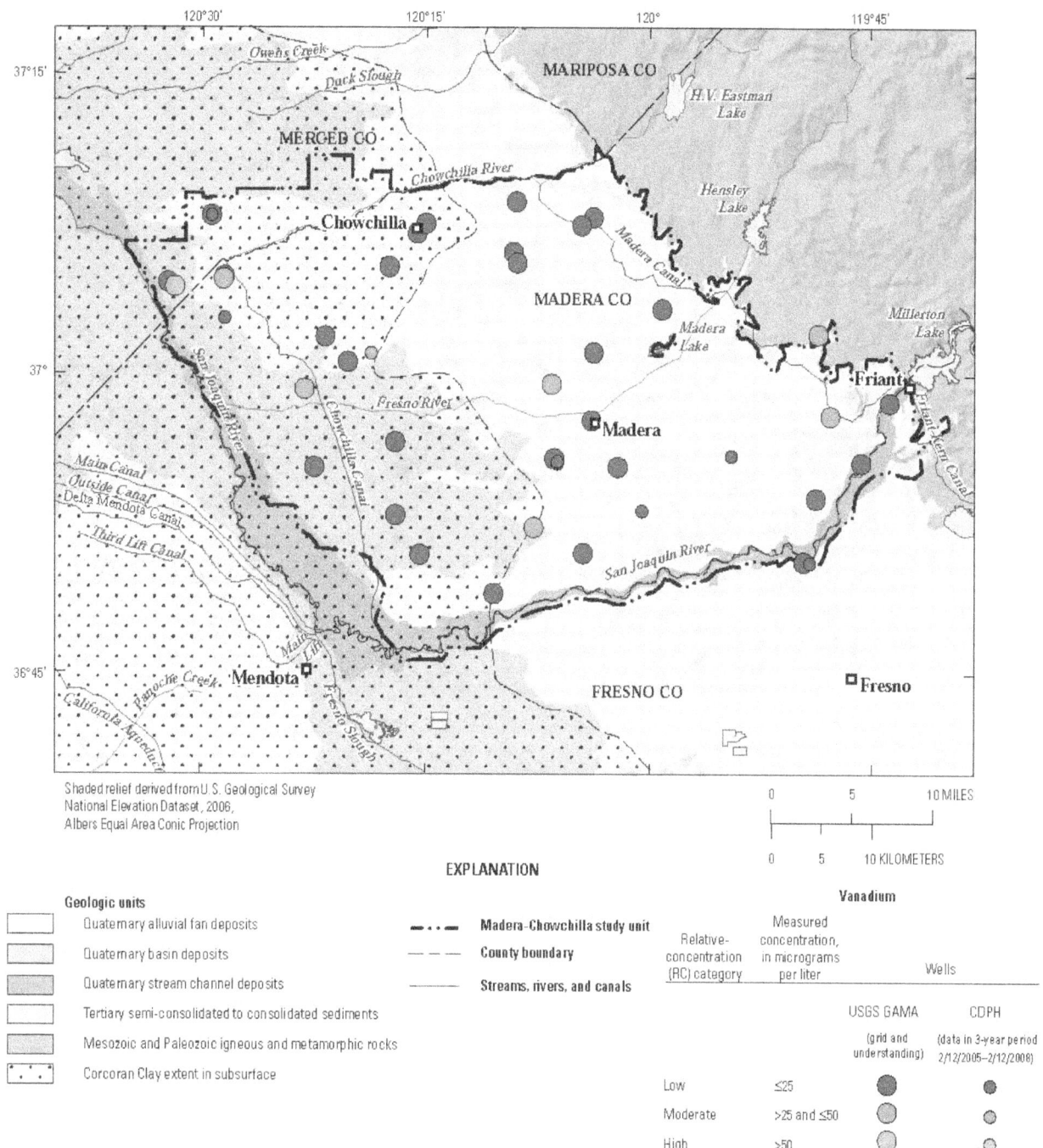

Figure 18. Concentrations of vanadium in USGS-GAMA wells and the most recent analysis during February 12, 2005– February 12, 2008, for CDPH wells, Madera-Chowchilla study unit, California GAMA Priority Basin Project.

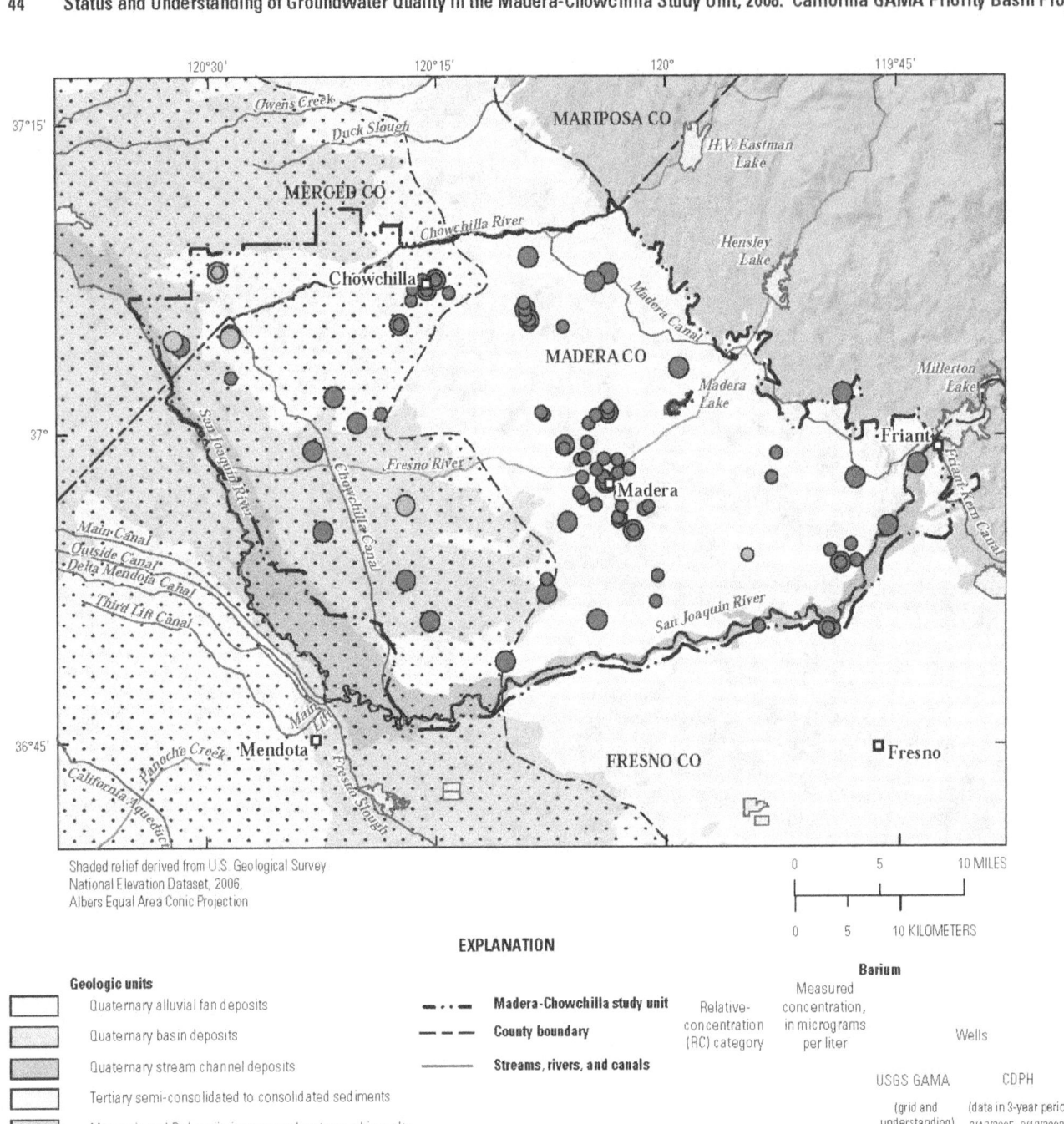

Figure 19.   Concentrations of barium in USGS-GAMA wells and the most recent analysis during February 12, 2005–February 12, 2008, for CDPH wells, Madera-Chowchilla study unit, California GAMA Priority Basin Project.

Factors Affecting Arsenic

Higher arsenic concentrations were significantly correlated with wells classified as deep wells (table 10A), but were not significantly correlated with either well depth or depth to top of perforations (table 10B; fig. 15B). Arsenic concentrations were positively correlated with pH and negatively correlated with dissolved oxygen (table 10B; fig. 20).

Previous investigations of arsenic in the San Joaquin Valley and literature reviews have attributed elevated arsenic concentrations in groundwater to two mechanisms (Welch and others, 2000, 2006; Belitz and others, 2003; Stollenwerk, 2003; Izbicki and others, 2008; Jurgens and others, 2008; Landon and others, 2010). One is the release of arsenic from dissolution of iron or manganese oxyhydroxides under iron- or manganese-reducing conditions. The other is desorption from (or inhibition of sorption to) aquifer sediments under oxic conditions with increasing pH. In the Madera-Chowchilla study unit, evidence for the first mechanism includes association of high RCs of arsenic with manganese-reducing conditions. The only USGS-GAMA sample with dissolved oxygen less than 0.5 mg/L had manganese-reducing conditions (table A5) and had the highest arsenic concentrations measured in USGS-GAMA samples (fig. 20A). This well was located in the northwest corner of the study unit. Elsewhere in the San Joaquin Valley, groundwater with high and moderate RCs of arsenic commonly occurs in the axial trough of the Valley, resulting in significant correlation between arsenic and lateral position (Belitz and others, 2003; Bennett and others, 2010; Landon and others, 2010). Arsenic was not significantly correlated with lateral position in the Madera-Chowchilla study unit because, unlike elsewhere in the San Joaquin Valley, the axial trough in the Madera-Chowchilla study unit is not dominated by anoxic oxidation-reduction conditions (fig. 11A).

Evidence for the second mechanism includes association of high and moderate RCs of arsenic with high pH in some wells (fig. 20B). Although the correlation between arsenic and pH was statistically significant (table 10B), there were many samples with pH values greater than 7.5 that had low RCs of arsenic. Of the six oxic samples with arsenic concentrations greater than 5 µg/L, only two have pH ≥ 8.0, and of the five samples with pH ≥ 8.0, only two have arsenic concentrations greater than 5 µg/L, suggesting pH-controlled sorption processes alone are not a sufficient explanation for the distribution of elevated arsenic concentrations in Madera-Chowchilla study unit groundwater.

The spatial distribution of elevated arsenic suggests that the composition of aquifer sediments is also a controlling factor. In the Madera-Chowchilla study unit, six of the seven USGS-GAMA samples with arsenic concentrations greater than 5 µg/L were from sites within 5 miles of the Chowchilla River along the northern margin of the study unit at lateral positions ranging from 0.05 to 0.92 (figs. 8, 17). Wells with high RCs of arsenic and lateral positions ranging from 0.1 to 0.8 also were close to the Chowchilla River in the Central Eastside study unit immediately to the north of the Madera-Chowchilla study unit (Landon and others, 2010).

Sediments in the Chowchilla River alluvial fan are lithologically different from those in the alluvial fans from some of the larger rivers (Kings, San Joaquin, Merced, Tuolumne, Stanislaus, and Mokelumne) that compose the rest of the aquifer matrix in the eastern San Joaquin Valley (Weissmann and others, 2005). The Chowchilla River watershed is confined to the lower elevations of the Sierra Nevada foothills in an area where metamorphic rocks are abundant (Saucedo and others, 2000). In contrast, the watersheds of the large rivers, such as the San Joaquin River, extend to higher elevations in the Sierra Nevada and are dominated by granitic rocks. The difference in lithology of the sediment source for the Chowchilla River compared to the sediment sources for the other rivers is reflected in the compositions of soils formed on top of the fans (Weissmann and others, 2005).

The arsenic is weathered from minerals in the sediments derived from the source rocks in the foothills of the Sierra Nevada. In the Sierra Nevada, sulfide minerals, such as pyrite, that are associated with metamorphic rocks in the foothills generally are more abundant and contain more arsenic than minerals associated with the granitic rocks (Izbicki and others, 2008). Thus, the sediments of the Chowchilla River alluvial fan may contain more arsenic that those of the San Joaquin River alluvial fan. In oxic groundwater, the sulfide minerals may be oxidized, releasing their arsenic into the groundwater.

The association between higher arsenic concentrations and deep wells suggests that residence time was also an important factor. Among the wells located close to the Chowchilla River, high and moderate RCs of arsenic primarily occurred in wells classified as deep. Shallow wells had low RCs of arsenic. The deep wells with high and moderate RCs of arsenic had mixed or pre-modern groundwater ages (fig. 15B). Arsenic was not significantly associated with groundwater age (table 10A) because only the deep wells near the Chowchilla River had elevated arsenic concentrations.

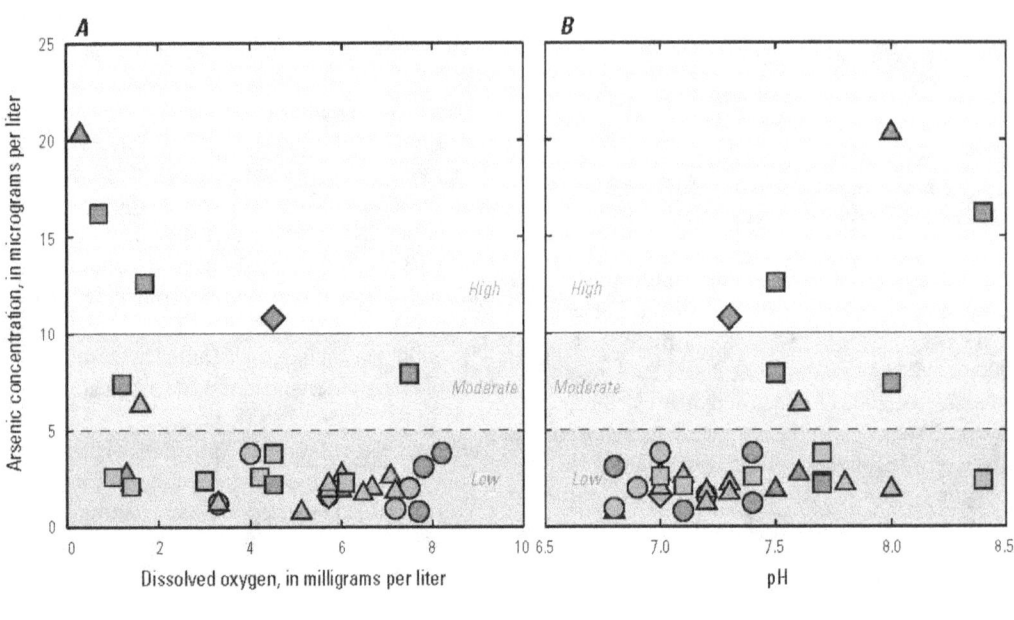

Figure 20.   Relations of (A) dissolved oxygen and (B) pH values to arsenic concentration and distance from Chowchilla River, Madera-Chowchilla study unit, 2008, California GAMA Priority Basin Project.

Factors Affecting Vanadium

Vanadium concentrations were not significantly correlated with any of the explanatory factors tested (tables 10A,B). Wright and Belitz (2010) investigated the regional distribution of vanadium in groundwater in California and reported that high vanadium concentrations were almost always associated with oxic and alkaline groundwater conditions. This correlation was not observed in groundwater samples from the Madera-Chowchilla study unit. The groundwater sample that contained a high RC of vanadium had a pH value of 8.4 and contained 0.7 mg/L of DO; however, pH values of the six groundwater samples with moderate RCs of vanadium ranged from 7.0 to 8.4. Vanadium was not correlated with well depth factors (fig. 15C), and was not correlated with groundwater age, although moderate and high RCs occurred only in groundwater with mixed or pre-modern ages.

Radioactive Constituents

The MCL-US (15 pCi/L) for gross alpha particle activity applies to adjusted gross alpha particle activity, which is equal to the measured gross alpha particle activity minus uranium activity (U.S. Environmental Protection Agency, 2000). Data collected by USGS-GAMA and data compiled in the CDPH database are reported as gross alpha particle activity without correction for uranium activity. Gross alpha particle activity is used as a screening tool to determine whether additional radioactive constituents must be analyzed (California Department of Public Health, 2012). For regulatory purposes, analysis of uranium is only required if gross alpha particle activity is greater than 15 pCi/L; therefore, the CDPH database contains more data for gross alpha particle activity than for uranium. As a result, it is not always possible to calculate adjusted gross alpha particle activity. For this reason, gross alpha particle activity data without correction for uranium are the primary data used in the status assessments made by USGS-GAMA for the Priority Basin Project. Examination of data from samples having USGS-GAMA data for uranium and gross alpha particle activity indicated that, in the absence of data for uranium, uncorrected gross alpha particle activity data likely provides a more accurate estimate of the aquifer-scale proportions for uranium and radioactive constituents as a class, than does adjusted gross alpha particle activity (Miranda Fram, USGS California Water Science Center, written commun., 2012).

USGS-GAMA reports data for gross alpha particle activity counted 72 hours and 30 days after sample collection. Gross alpha particle activity in a groundwater sample may change with time after sample collection due to radioactive decay and ingrowth (activity may increase or decrease depending on sample composition and holding time) (Arndt, 2010). Data from the 72-hour counts are used in this report.

Most data for uranium in the CDPH database are reported as activities in units of pCi/L, and the majority of uranium data gathered by USGS-GAMA are reported as concentrations in units of micrograms per liter. The factor used to convert uranium mass concentration to uranium activity depends on the isotopic composition of the uranium (U.S. Environmental Protection Agency, 2000). This report uses a conversion factor of 0.79.

Radioactive constituents with health-based benchmarks were present at high RCs in 20% of the primary aquifer system, and at moderate and low RCs in 3.3% and 77%, respectively (table 9A). Gross alpha particle activity and uranium activity were the radioactive constituents detected at high or moderate RCs in the primary aquifer system (table 8).

Gross alpha particle activity was detected at high RCs in 20% of the primary aquifer system, and uranium activity was detected at high RCs in 17%; moderate RCs of both were present in 3.3% of the primary aquifer system (table 8). Of the six grid wells with high RCs of gross alpha particle activity, five had high RCs of uranium activity and one had moderate RC. A high RC for radium was reported in the CDPH database for one well during the 3-year period February 2005–February 2008 (table 5), but the high value was not from the most recent sample; therefore, no high RC is reported using the spatially weighted approach (table 8).

Factors Affecting Uranium and Gross Alpha Particle Activity

Groundwater with high and moderate RCs of uranium activity was primarily from wells located in the western portion of the study unit (fig. 21). Uranium and gross alpha particle activity were closely correlated (Spearman's test; p<0.001, rho=0.73), and the patterns of significant correlations with potential explanatory factors were similar for the two constituents (tables 10A,B). The following discussion is limited to factors affecting uranium.

Geochemical conditions, groundwater age, and depth were the most significant explanatory factors related to uranium activities. Uranium activities were significantly greater in modern groundwater than in pre-modern groundwater, and in wells classified as shallow or mixed than in wells classified as deep (table 10A). Uranium was significantly negatively correlated with depth to the top of perforation (table 10B; fig. 15D), and positively correlated with dissolved oxygen concentration (table 10B).

This significant association between higher uranium activities, modern-age groundwater, oxic conditions, and shallow depths in the Madera-Chowchilla study unit is the same as the pattern observed in the eastern San Joaquin Valley as a whole (Jurgens and others, 2009a). Jurgens and others (2009a) attributed the elevated uranium in shallow modern-age groundwater to enhanced desorption of uranium from soil and aquifer sediments by recharge of water used for irrigation that has high bicarbonate concentrations. The bicarbonate is derived from biological production of carbon dioxide in the soil zones of irrigated landscapes. The groundwater budget in the eastern San Joaquin Valley is dominated by irrigation recharge and pumping (fig. 4; Faunt, 2009), which results in transport of irrigation recharge to depths in the aquifer system tapped by public-supply wells.

In the Madera-Chowchilla study unit, uranium activity had a significant positive correlation with bicarbonate (rho=0.800, p<0.001) (fig. 22), which is consistent with desorption of uranium from aquifer sediments by complexation with dissolved bicarbonate. Uranium also had a significant positive correlation with calcium (rho=0.678, p<0.001) and with Fract-CaMg (table 10B), which may indicate that part of the increase in bicarbonate is caused by dissolution of soil and sedimentary calcite.

Although all of the high RCs and most of the moderate RCs of uranium occurred in wells located in the western part of the study unit, uranium was not significantly correlated with lateral position (table 10A). However, the statistical test was made using only the grid wells; uranium had a significant negative correlation with lateral position when the grid and understanding wells were considered (Spearman's test; p=0.020, rho=−0.39).

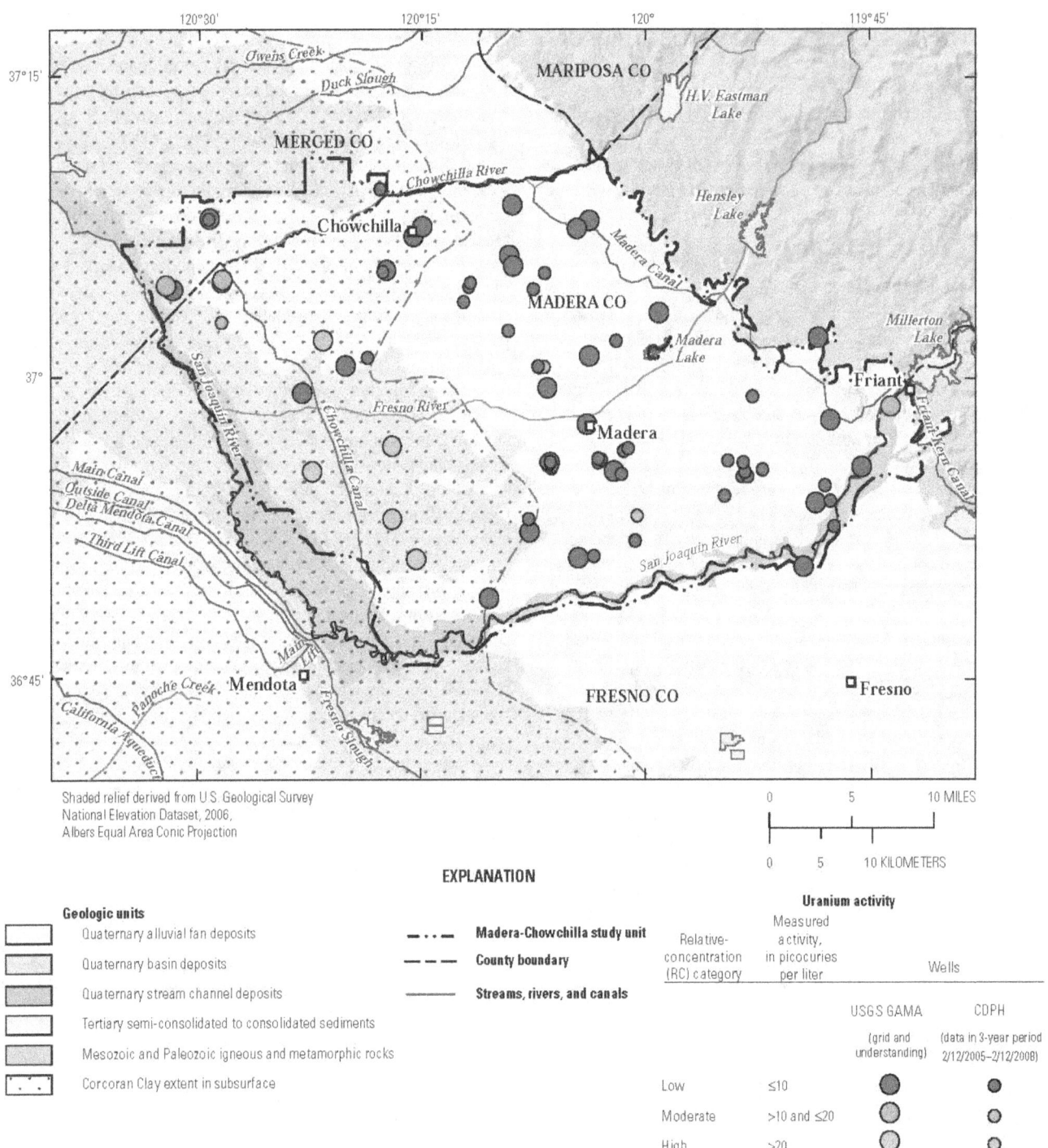

Figure 21.   Activities of uranium in USGS-GAMA wells and the most recent analysis during February 12, 2005–February 12, 2008, for CDPH wells, Madera-Chowchilla study unit, California GAMA Priority Basin Project.

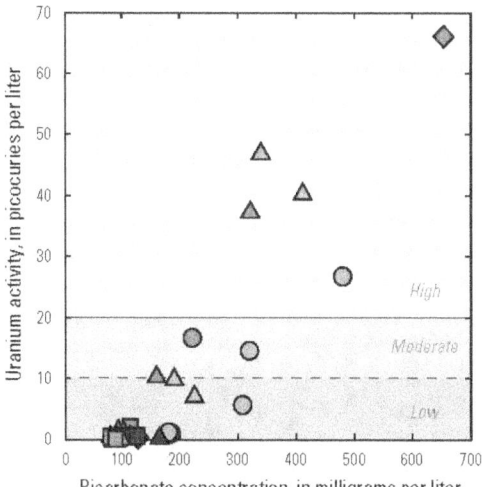

EXPLANATION

Groundwater age classification			Well depth classification
Modern	Mixed	Pre-modern	
◯	◉		Shallow (Entire perforated interval ≤ 280 feet)
△	▲	▲	Mixed (Top of perforated interval ≤ 280 feet; Bottom of perforated interval > 280 feet)
	■	■	Deep (Entire perforated interval > 280 feet)
	◇		Unknown

Figure 22. Relation of bicarbonate concentrations and uranium activities, Madera-Chowchilla study unit, 2008, California GAMA Priority Basin Project.

Constituents with SMCL Benchmarks

Constituents with SMCL benchmarks were present at high RCs in 6.7% of the primary aquifer system, at moderate RCs in 17%, and at low RCs in 77% (table 9A). The constituent most commonly responsible for the high RCs was TDS, which was present at high RCs in 6.7% of the primary aquifer system (table 8). Chloride and manganese were each present at high RCs in 3.3% of the primary aquifer system, and moderate RCs of manganese were observed in 3.3% of the primary aquifer system. Nearly all of the wells with high and moderate RCs of TDS were located in the western portion of the study unit (fig. 23). Two of the wells with high RCs of TDS also had high RCs of chloride, and one of these wells also had a high RC of manganese. These two wells were located in the northwestern corner of the study unit.

All detections of iron in grid and understanding wells had low RCs, thus the grid-based calculation yielded a high aquifer-scale proportion of 0%. However, iron was reported at high RCs in 13 wells in the CDPH database, and the resulting spatially weighted high aquifer-scale proportion, 7.0%, was outside of the 90% confidence interval for the grid-based high aquifer-scale proportion (table 8). The 13 CDPH wells with high RCs of iron were distributed throughout the study unit, and of those 13, only 4 also had high RCs of manganese. Because iron reduction typically occurs at lower oxidation-reduction potentials than manganese reduction (Appelo and Postma, 2005; McMahon and Chapelle, 2008), one would not expect to find groundwater with elevated iron concentrations without elevated manganese concentrations. As discussed in appendix B, these results may be due to the sampling methods typically used for samples collected for analysis of trace elements. Anoxic conditions, including elevated iron and manganese concentrations, have been found historically in the southwestern portion of the study unit (Mitten and others, 1970). Two of the wells were also sampled by USGS-GAMA (MADCHOW-05, -09). Specific conductance in MADCHOW-05 was 149 microsiemens per centimeter at 25 degrees Celsius (µS/cm at 25°C) in November 2007 (CDPH) and 272 µS/cm at 25°C in April 2008 (USGS-GAMA), and for MADCHOW-09, specific conductance was 310 µS/cm at 25°C in January 2006 and 760 µS/cm at 25°C in April 2008 (USGS-GAMA). These differences may indicate that water quality has changed over time and that high iron concentrations may not be representative of the current water-quality conditions.

Factors Affecting Total Dissolved Solids

Groundwater age, depth, and geochemical conditions were the most significant explanatory factors related to TDS concentrations. The pattern of correlations between TDS and potential explanatory factors was similar to the patterns shown by nitrate and uranium (tables 10A,B). Similar to nitrate and uranium concentrations, TDS concentrations were significantly higher in modern groundwater as compared to pre-modern groundwater, in wells with shallower depths to the top of perforations compared with wells perforated at deeper depths, and in wells classified as above/across the Corcoran Clay compared with other positions relative to the Clay (tables 10A,B; fig. 15E). TDS concentrations had significant positive correlations with uranium activities (rho = 0.593, $p < 0.001$) and nitrate concentrations (rho = 0.566, $p < 0.001$).

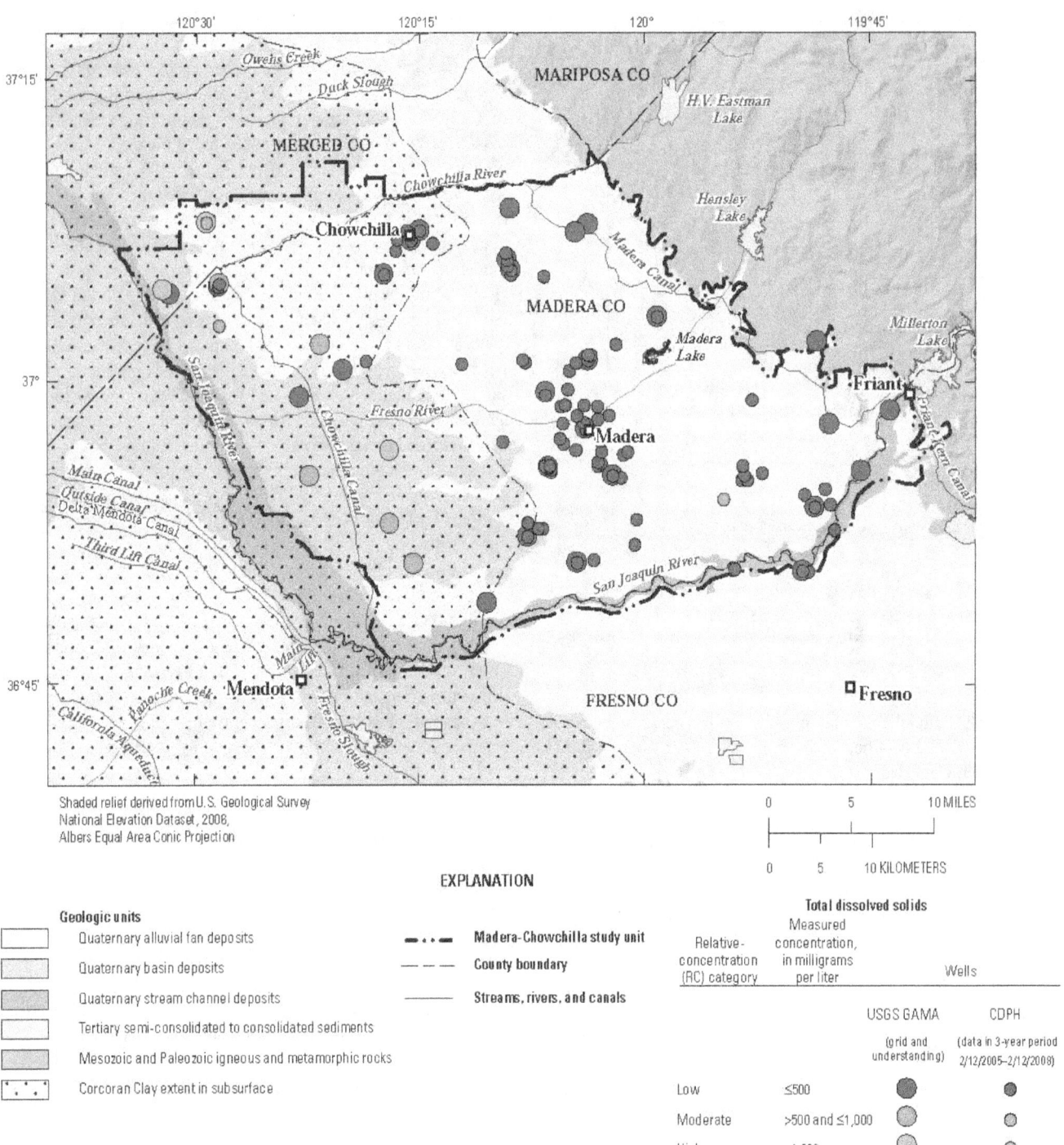

Figure 23.   Concentrations of total dissolved solids (TDS) in USGS-GAMA wells and the most recent analysis during February 12, 2005–February 12, 2008, for CDPH wells, Madera-Chowchilla study unit, California GAMA Priority Basin Project.

Together, these correlations suggest that increases in TDS, nitrate, and uranium may be related to similar processes. The higher concentrations of TDS (and nitrate and uranium) in shallower groundwater imply greater loading of dissolved constituents from the surface to groundwater in recent decades. This greater loading may be the result of several factors, including increases in recharge and fluctuation of water levels, changes in soil and soil pore-water chemistry caused by historical changes in land use, use of fertilizers, soil amendments and other chemicals at the land surface, and increases in consumptive water use by vegetation.

Previous studies of eastern San Joaquin Valley groundwater have found negative correlations between TDS and lateral position (Mendenhall and others, 1916; Bertoldi and others, 1991; Bennett and others, 2010; Landon and others, 2010). Groundwater in the eastern alluvial fans typically has lower TDS concentrations than groundwater in the basin area. TDS was not correlated with lateral position (grid wells only) in the Madera-Chowchilla study unit (table 10B); however, for both grid and understanding wells, TDS had a significant negative correlation with lateral position (rho = –0.35, p=0.040) because the understanding wells typically were shallower than the grid wells (fig. 7). For the 24 wells classified as shallow or mixed depth, TDS was significantly negatively correlated with lateral position (rho = –0.71, p<0.001). The increase in TDS towards the center of the Valley may reflect a variety of natural and anthropogenic processes which include historical groundwater discharge and evapotranspiration patterns, irrigation return and irrigation recycling, addition of salts from anthropogenic activities at or near the land surface, rock/water interaction along regional groundwater flow paths, and upwelling of more saline groundwater influenced by interactions with deep marine or lacustrine sediments near the valley trough.

Detailed analysis of the processes accounting for increases in TDS is beyond the scope of this report, although the relations between TDS concentrations and geochemical characteristics of the groundwater may provide some insight into the sources of salinity. Major ion groundwater chemistry is represented on a Piper diagram, which shows the proportions of the major cations (calcium, magnesium, and sodium plus potassium) and the major anions (bicarbonate, sulfate, and chloride) on a charge-equivalent basis (fig. 24; Piper, 1944; Hem, 1992). The majority of the Madera-Chowchilla groundwater samples have calcium or calcium plus sodium as the primary cations and bicarbonate as the primary anion. Samples from four wells have sodium as the primary cation and bicarbonate as the primary anion, and samples from three wells have chloride or chloride plus bicarbonate as the primary anions. This distribution of geochemical types is similar to that observed in the Central Eastside study unit (Landon and others, 2010) and in the Northern San Joaquin Valley (Izbicki and others, 2006; Bennett and others, 2010). Geographically, groundwater in the eastern alluvial fans typically has calcium plus sodium as the

primary cations and bicarbonate as the primary anion, whereas groundwater in the basin area has a variety of compositions, with calcium and/or sodium as the primary cations and bicarbonate and/or chloride as the primary anions (Bertoldi and others, 1991).

The distribution of groundwater samples shown on the lower left portion of the Piper diagram shows a distinction between groundwater chemistry in wells with lower TDS and wells with higher TDS. Wells with higher TDS (>500 mg/L) yield calcium-dominated bicarbonate water that typifies the shallow and mixed depth wells. Wells with lower TDS (<500 mg/L) yield sodium-potassium bicarbonate water that mainly typifies the deeper wells.

Although TDS concentrations had a significant positive correlation with dissolved oxygen and a significant negative correlation with pH (table 10B), the only well with anoxic groundwater (MADCHOW-12) also had a high RC of TDS (1,070 mg/L) and a pH of 8.0. MADCHOW-12 is located in the the northwest corner of the basin area of study unit and is perforated entirely below the Corcoran Clay (table A2; figs. 2, 3). The primary cation in the sample from MADCHOW-12 is sodium, and the primary anion is chloride (fig. 24). The chloride-to-iodide ratio in this sample is low compared to the other samples from the Madera-Chowchilla study unit and is in the range of chloride-to-iodide ratios of groundwater affected by interactions with marine sediments (Izbicki and others, 2006). One sample from a well perforated entirely below the Corcoran Clay in the basin area of the Central Eastside GAMA study unit located immediately north of the Madera-Chowchilla GAMA study unit had a similar anion composition (Landon and others, 2010). Landon and others (2010) concluded that this sample may represent upwelling of deeper saline waters from the marine sedimentary deposits beneath the continental deposits that compose the freshwater aquifer system as a result of upward hydraulic gradients at the distal end of the regional groundwater flow system.

With the exception of MADCHOW-12, all of the samples with high or moderate RCs of TDS were perforated above and (or) across the Corcoran Clay (fig. 15E) and were oxic. All samples with high or moderate RCs of TDS had mixed or modern groundwater ages (fig. 15E). TDS had a significant positive correlation with Fract-CaMg (table 10B). Groundwater samples with low RCs of TDS generally have cation compositions closer to the sodium plus potassium apex (lower Fract-CaMg) of a Piper diagram, and groundwater samples with moderate or high RCs of TDS lie furthest from the sodium plus potassium apex (higher Fract-CaMg) (fig. 24). One possible mechanism for this elevated TDS and Fract-CaMg in modern, shallow groundwater may be enhanced dissolution of calcite from soils by irrigation recharge, similar to the mechanism responsible for increased uranium concentrations in modern, shallow groundwater in the eastern San Joaquin Valley (Jurgens and others, 2009a).

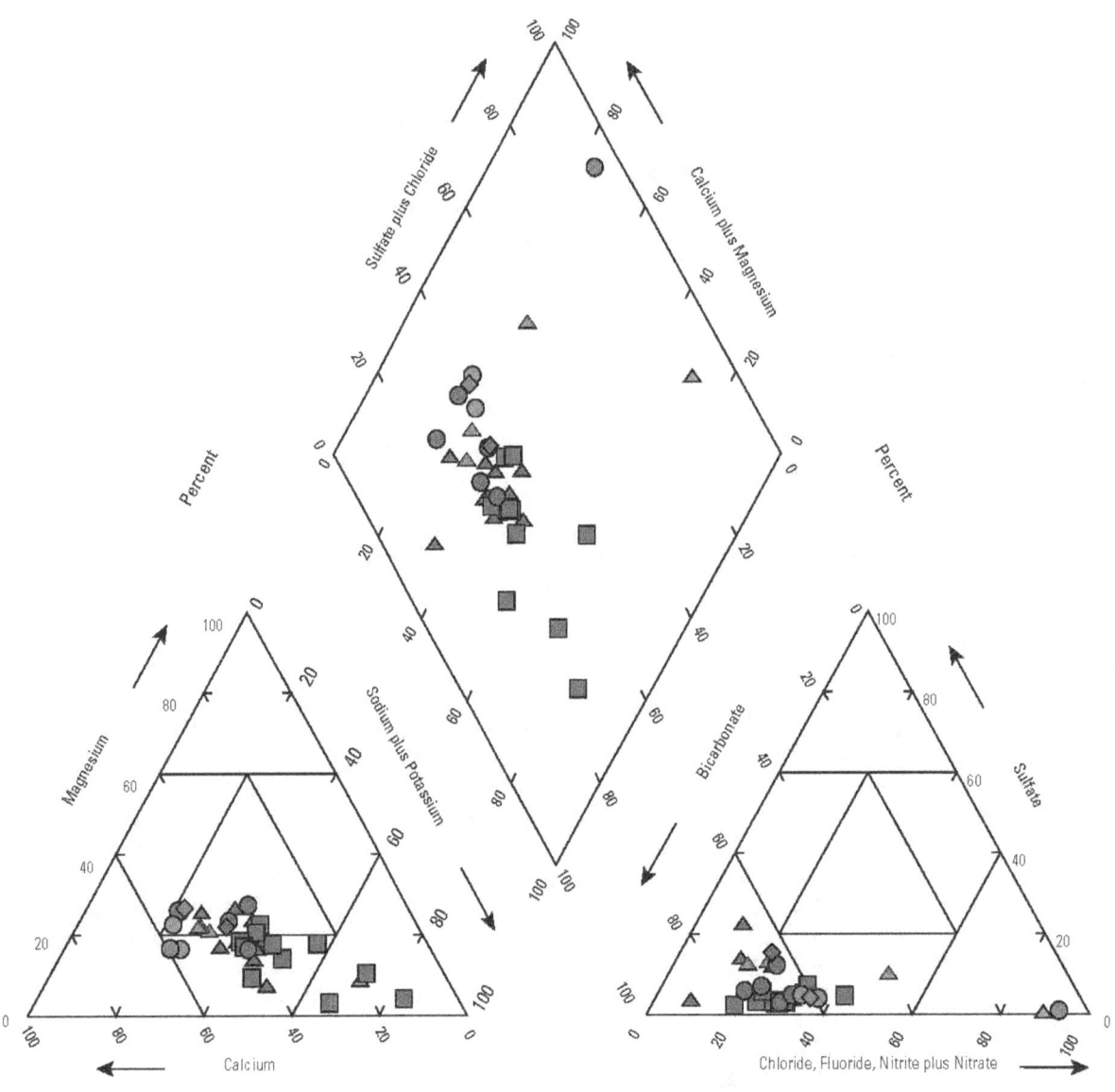

Figure 24.   Major-ion composition and total dissolved solids concentrations for groundwater samples from grid and understanding wells, 2008, Madera-Chowchilla study unit, California GAMA Priority Basin Project.

Factors Affecting Manganese

Manganese concentrations were significantly higher in pre-modern groundwater than in modern or mixed-age groundwaters, and manganese concentrations were positively correlated with urban land use, pH, and depth to the top of perforations (tables 10A, B). Manganese concentrations were not correlated with dissolved oxygen concentrations (table 10B) possibly because the only sample with a high RC for manganese (MADCHOW-12) was also the only sample with dissolved oxygen concentration less than 0.5 mg/L (table A5). All 10 samples with manganese concentrations greater than 3 µg/L were from wells located within 5 miles of the Chowchilla River, similar to the areal distribution of wells with moderate and high RCs of arsenic (fig. 17).

Organic and Special-Interest Constituents

For this report, the organic and special-interest constituents are organized by constituent class: pesticides, VOCs, and special interest. Volatile organic compounds (VOCs) are present in paints, solvents, fuels, fuel additives, refrigerants, fumigants, and disinfected water, and are characterized by their tendency to evaporate. VOCs typically persist longer in groundwater than in surface water because groundwater is isolated from the atmosphere. Pesticides include herbicides, insecticides, and fumigants, and are applied to crops, gardens, lawns, around buildings, and along roads to help control unwanted vegetation (weeds), insects, fungi, and other pests in agricultural, urban, and suburban settings. Fumigants can be classified as VOCs because of their tendency to evaporate, but also can be classified as pesticides because they are used to control pests. Fumigants are of interest in California because of their historical use on vineyards and orchards. The constituents of special interest group includes two chemically unrelated constituents (perchlorate and *N*-nitrosodimethylamine) that are of interest in California because they have recently been detected in groundwater because of advances in analytical methods.

USGS-GAMA included analysis of a large number of organic constituents, many of which are not subject to any regulation in drinking water, and used analytical methods with lower detection limits than required for regulatory sampling. In the Madera-Chowchilla study unit, however, the majority of organic constituents detected were already subject to drinking-water regulations. Of the 134 pesticides and pesticide degradates analyzed, 12 were detected in at least one well (table 7B). Seven of those 12 had regulatory or non-regulatory health-based benchmarks. Of the five pesticide constituents detected with no benchmarks, three were degradates of parent compounds with benchmarks (table 4). Of the 10 fumigants analyzed, 4 were detected in at least one well, and all 4 had health-based benchmarks. Of the 75 other VOCs analyzed,

8 were detected in at least one of the wells, and all have regulatory health-based benchmarks. Of the two special-interest constituents, both were detected, and both have health-based benchmarks. Of the 118 organic and special-interest constituents with no health-based benchmarks analyzed in this study unit, 5 were detected in groundwater.

Figure 25 summarizes the study-unit detection frequencies and maximum RCs for organic and special-interest constituents detected in the Madera-Chowchilla study unit. The fumigant 1,2-dibromo-3-chloropropane (DBCP), the solvent tetrachloroethene (PCE), and perchlorate were selected for additional evaluation in the *status assessment* because they had maximum RCs greater than 0.1 and study-unit detection frequencies greater than or equal to 10%. The fumigant 1,2-dibromoethane (EDB) was selected for additional evaluation because it had a maximum RC greater than 0.1. Atrazine, simazine, diuron, chloroform, and 1,2,3-TCP were selected because they had study-unit detection frequencies greater than or equal to 10% (figs. 25, 26; table 8). Eleven other organic constituents with health-based benchmarks were detected in grid wells at RCs less than 0.1 and had study-unit detection frequencies less than 10% (fig. 25; table 4).

Aquifer-scale proportions for individual organic and special-interest constituents are listed in table 8, and results for organic constituent classes are listed in table 9B. For any organic constituent having health-based benchmarks (pesticides and VOCs), 10% of the primary aquifer system, on an areal basis, had high RCs of at least one constituent, 3.3% had moderate values, 40% had detections of organic constituents at low values, and 47% had no detections of organic constituents (table 9B).

Pesticides

Herbicides were not detected at high or moderate RCs in the Madera-Chowchilla study unit (tables 8, 9B); the maximum RC detected was 0.039 (atrazine) (figs. 25, 26). Twenty-three percent of grid wells sampled (7 of 30) had at least one herbicide detected (Shelton and others, 2009). Simazine was detected in 20% of the grid wells, and diuron and atrazine were each detected in 10% of the grid wells (figs. 25, 26). All of the grid well samples containing diuron also contained 3,4-dichloroaniline, a degradation product, and all of the grid well samples containing atrazine also contained the degradation product deethylatrazine. Deisopropyl atrazine, another a degradation product of atrazine, was detected in the sample containing the highest concentrations of atrazine and deethylatrazine. The degradation products of atrazine and diuron were detected in more samples than were these two parent compounds (Shelton and others, 2009). Benchmarks have not been established for these degradation products; thus, they are not included in the *status assessment*.

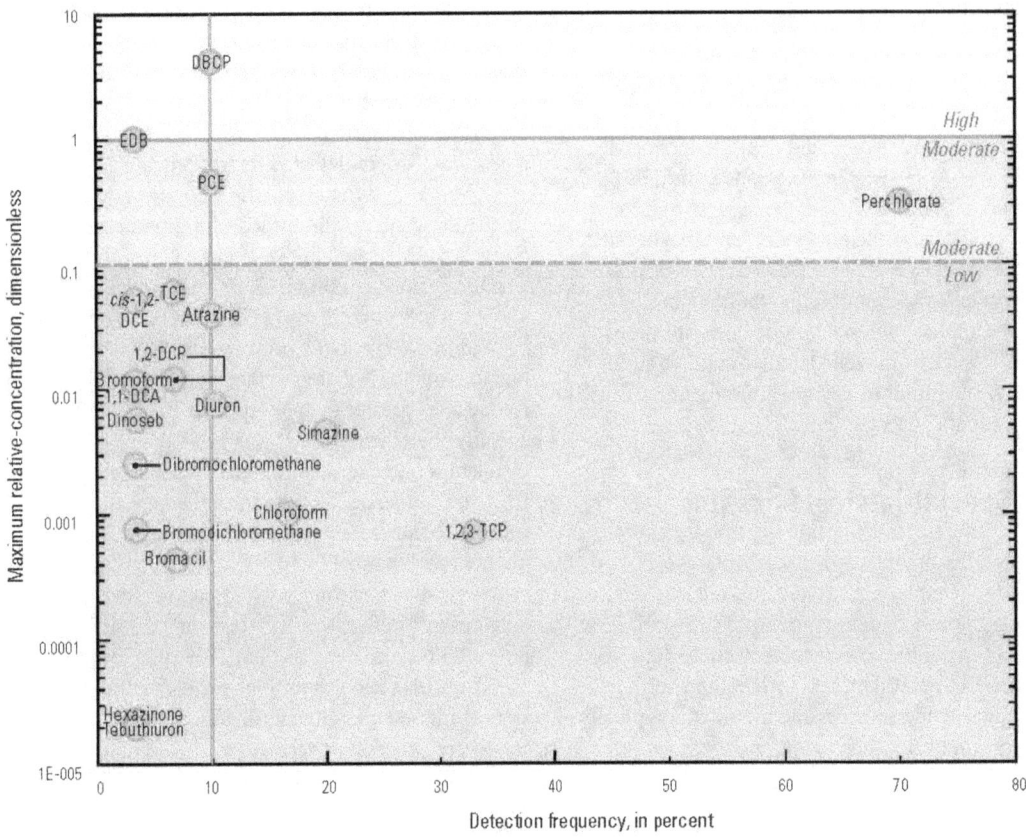

EXPLANATION

PCE   **Constituents with analyses in >20 grid wells and wells are spatially representative—**
Name and center of symbol is location of data unless indicated by following location line: ●———⌐

Abbreviations

cis-1,2-DCE, cis-1,2-dichloroethene; DBCP, 1,2-dibromo-3-chloropropane; EDB, 1,2-dibromoethane; 1,1-DCA, 1,1-dichloroethane; 1,2-DCP, 1,2-dichloropropane; PCE, tetrachloroethene; TCE, trichloroethene; 1,2,3-TCP, 1,2,3-trichloropropane.

Figure 25.   Detection frequency and maximum relative-concentrations for organic and special-interest constituents detected in grid wells, 2008, Madera-Chowchilla study unit, California GAMA Priority Basin Project.

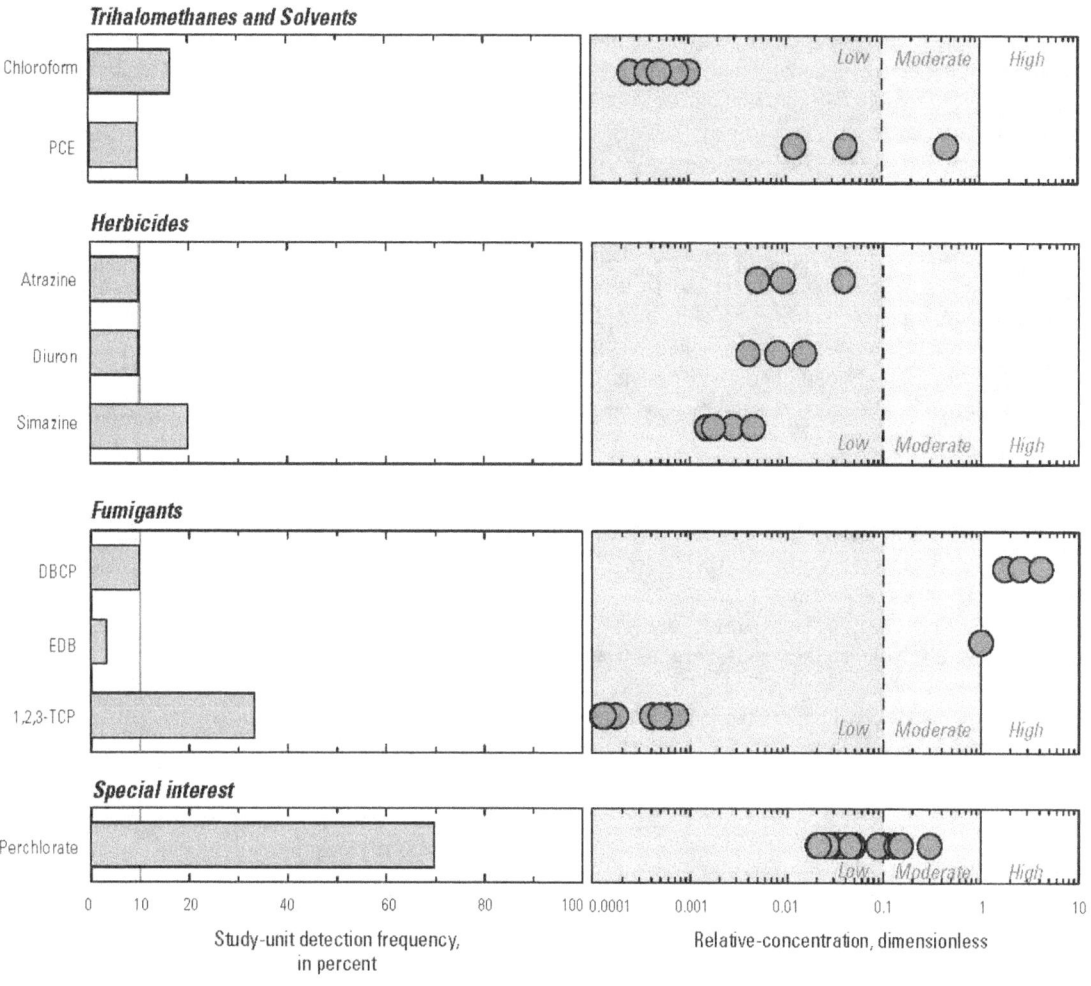

Figure 26. Detection frequency and relative-concentrations of selected organic and special-interest constituents in grid wells, 2008, Madera-Chowchilla study unit, California GAMA Priority Basin Project.

EXPLANATION

Abbreviations

PCE, tetrachloroethene; DBCP, 1,2-dibromo-3-chloropropane; EDB, 1,2-dibromoethane; 1,2,3-TCP, 1,2,3-trichloropropane.

Detections of atrazine and deethylatrazine are the most common two-compound mixtures of pesticides detected in the groundwater sampled by the National Water-Quality Assessment (NAWQA) Program (Gilliom and others, 2006), and their co-occurrence may reflect the relatively high degree of persistence of atrazine in groundwater environments (Kolpin and others, 1998). Deethylatrazine, atrazine, and simazine were the most frequently detected pesticide compounds in groundwater in major aquifers across the United States (Gilliom and others, 2006). In California, simazine is most commonly used on orchards and vineyards and on rights-of-way for weed control. Diuron is most commonly used on rights-of-way, alfalfa for forage, and orchards and vineyards, and atrazine has historically been most commonly used on corn and rights-of-way (Kegley and others, 2008).

Herbicides were detected at low RCs in 9 of the 35 USGS-GAMA wells used in the *understanding assessment* (fig. 27A). Herbicides were not detected in any of the 69 CDPH wells for which data were reported for herbicides between February 12, 2005, and February 12, 2008 (fig. 27A). All of the detections in USGS-GAMA samples had concentrations that were lower than the reporting limits used in the CDPH database.

Insecticides were not detected in USGS-GAMA samples (not including the fumigants DBCP and EDB). No insecticide detections were reported in the CDPH database (not including the fumigants DBCP and EDB).

Factors Affecting Herbicides

Rather than considering atrazine, simazine, and diuron separately in the *understanding assessment*, concentrations of all herbicides with benchmarks were summed and treated as a constituent class.

Depth and groundwater age were the most significant factors affecting organic constituents (table 10A; fig. 28A). Herbicide concentrations were significantly greater in modern groundwater than in pre-modern or mixed groundwater, and in wells classified as shallow or mixed than in wells classified as deep. All of the wells with detections of herbicides had modern or mixed-age groundwater. Herbicide concentrations had a significant negative correlation with depth to the top of perforations, and herbicides were not detected in wells with the depths to top of perforation deeper than 240 ft below land surface.

Herbicides were correlated positively with dissolved oxygen and Fract-CaMg and negatively with pH and depth to top of perforations (table 10B). The correlations between herbicides and the geochemical explanatory factors likely result from the correlations between the geochemical explanatory factors and depth to top of perforations and shallow wells (tables 6A,B). Herbicide concentrations were not significantly correlated with land use (table 10B). The lack of correlation may reflect that the most frequently detected herbicides have agricultural and non-agricultural applications, or the lack of correlation may reflect the dominance of

agricultural land use in the study unit. Sixty-nine percent of the study unit has agricultural land use, and the areas of urban land use are small in area and surrounded by agricultural land use (figs. 5A, 6).

Fumigants

The proportion of the primary aquifer system with high RCs of fumigants was 10%. DBCP was the only fumigant (and the only organic constituent) present at high RCs, and all detections had high RCs, resulting in a high aquifer-scale proportion for DBCP of 10% (table 8; figs. 25, 26). One of the samples with a high RC of DBCP also had a moderate RC of another fumigant, EDB. The detection frequency of 1,2,3-TCP in grid wells was 33%, and all detections had low RCs (figs. 25, 26).

DBCP and EDB are soil fumigants that were used to control nematodes primarily in orchard and vineyards, but their usage was discontinued in 1977 and 1983, respectively (Domagalski, 1997; California State Water Resources Control Board, 2002; Kegley and others, 2008). 1,2,3-TCP was used in the manufacture of D-D (dichloropropane-dichloropropene mixture) (California Department of Public Health, 2009), a soil fumigant whose usage was discontinued in 1987 (Kegley and others, 2008). DBCP was the most frequently detected fumigant or pesticide detected in groundwater samples collected from the San Joaquin Valley during 1971–1988 (Domagalski, 1997) and in groundwater samples statewide in samples analyzed through 1999 (Troiano and others, 2001). Detection frequencies of DBCP in San Joaquin Valley groundwater have been higher than those reported in most other parts of the nation because of DBCP's historical use on orchards and vineyards in California (Dubrovsky and others, 1998; Zogorski and others, 2006). Vineyards in the San Joaquin Valley commonly have been located on soils with relatively coarse textures, and DBCP generally was applied by injection into the soils; these factors likely contributed to transport of DBCP to groundwater (Burow and others, 1998; 2000). Nationally, DBCP contributed to most of the concentrations of VOCs above MCLs or health-based screening levels (Zogorski and others, 2006).

In addition to the three USGS-GAMA wells with high RCs of DBCP, high or moderate RCs of DBCP were reported in three other wells in the CDPH database between February 12, 2005, and February 12, 2008, and low RCs were reported in another six CDPH wells (one of which was MADCHOW-18). For four of these nine CDPH wells, the detection of DBCP was not in the most recent sample analyzed, so these detections are not shown on fig. 27B. All nine CDPH wells and two of the three USGS wells with detections of DBCP were located south of the city of Madera (fig. 27B). EDB was detected in one USGS well and one CDPH well, both of which also had detections of DBCP and were located south of Madera. For one CDPH well, results for 240 analyses of DBCP and EDB were reported in the CDPH database between April 2005 and December 2007.

A. Herbicides

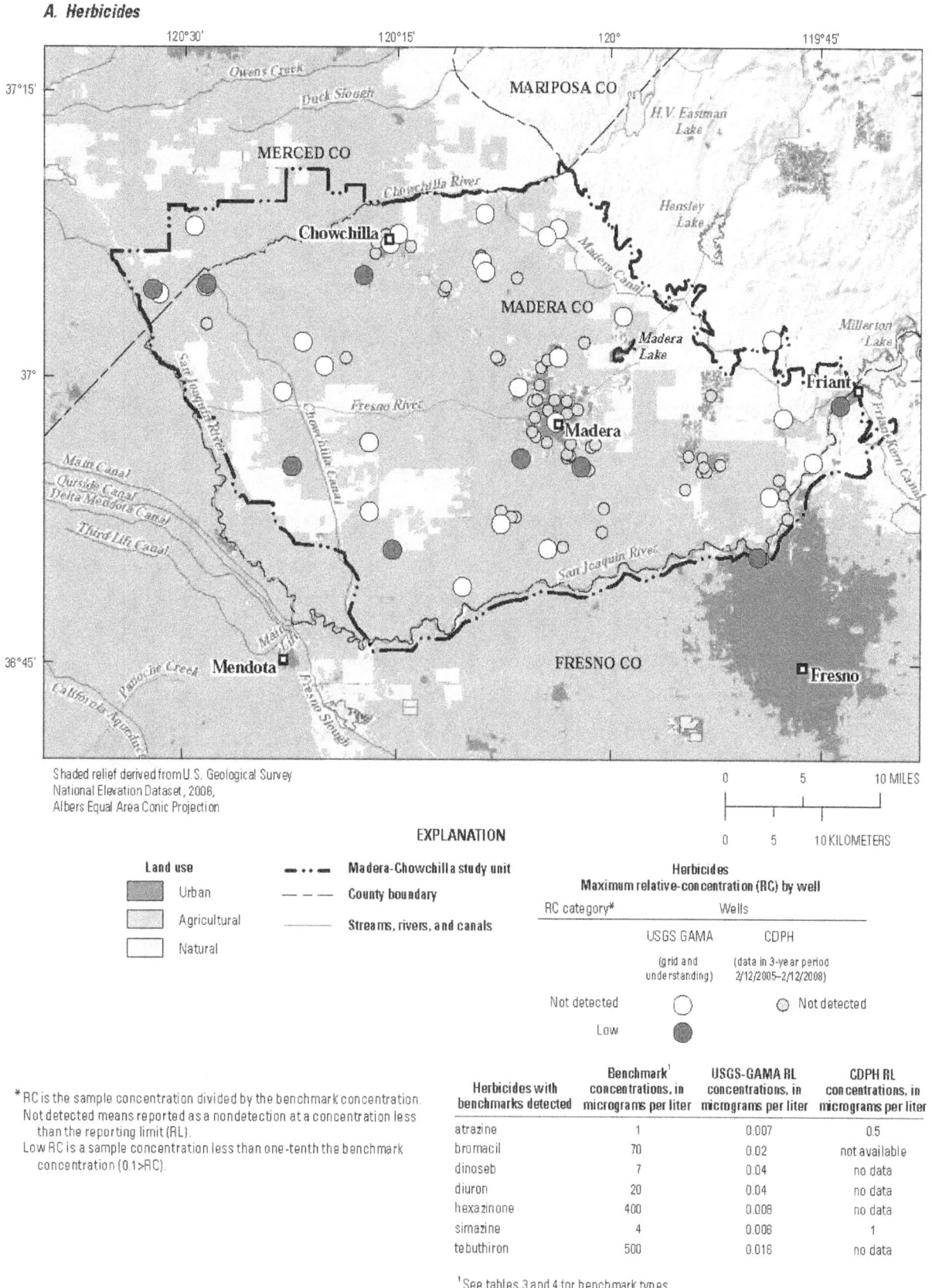

Shaded relief derived from U.S. Geological Survey
National Elevation Dataset, 2008,
Albers Equal Area Conic Projection

0   5   10 MILES

0   5   10 KILOMETERS

EXPLANATION

Land use
- Urban
- Agricultural
- Natural

— ·· — Madera-Chowchilla study unit
— — — County boundary
——— Streams, rivers, and canals

Herbicides
Maximum relative-concentration (RC) by well

RC category*	Wells	
	USGS GAMA (grid and understanding)	CDPH (data in 3-year period 2/12/2005–2/12/2008)
Not detected	○	◉ Not detected
Low	●	

*RC is the sample concentration divided by the benchmark concentration. Not detected means reported as a nondetection at a concentration less than the reporting limit (RL).
Low RC is a sample concentration less than one-tenth the benchmark concentration (0.1>RC).

Herbicides with benchmarks detected	Benchmark¹ concentrations, in micrograms per liter	USGS-GAMA RL concentrations, in micrograms per liter	CDPH RL concentrations, in micrograms per liter
atrazine	1	0.007	0.5
bromacil	70	0.02	not available
dinoseb	7	0.04	no data
diuron	20	0.04	no data
hexazinone	400	0.008	no data
simazine	4	0.006	1
tebuthiron	500	0.016	no data

¹See tables 3 and 4 for benchmark types.

Figure 27. Maximum relative-concentrations of selected constituents in organic constituent classes and perchlorate in USGS-GAMA wells and the most recent analysis during February 12, 2005–February 12, 2008, for CDPH wells, Madera-Chowchilla study unit, California GAMA Priority Basin Project: (A) herbicides, (B) fumigants, (C) trihalomethanes, (D) solvents, and (E) perchlorate.

B. Fumigants

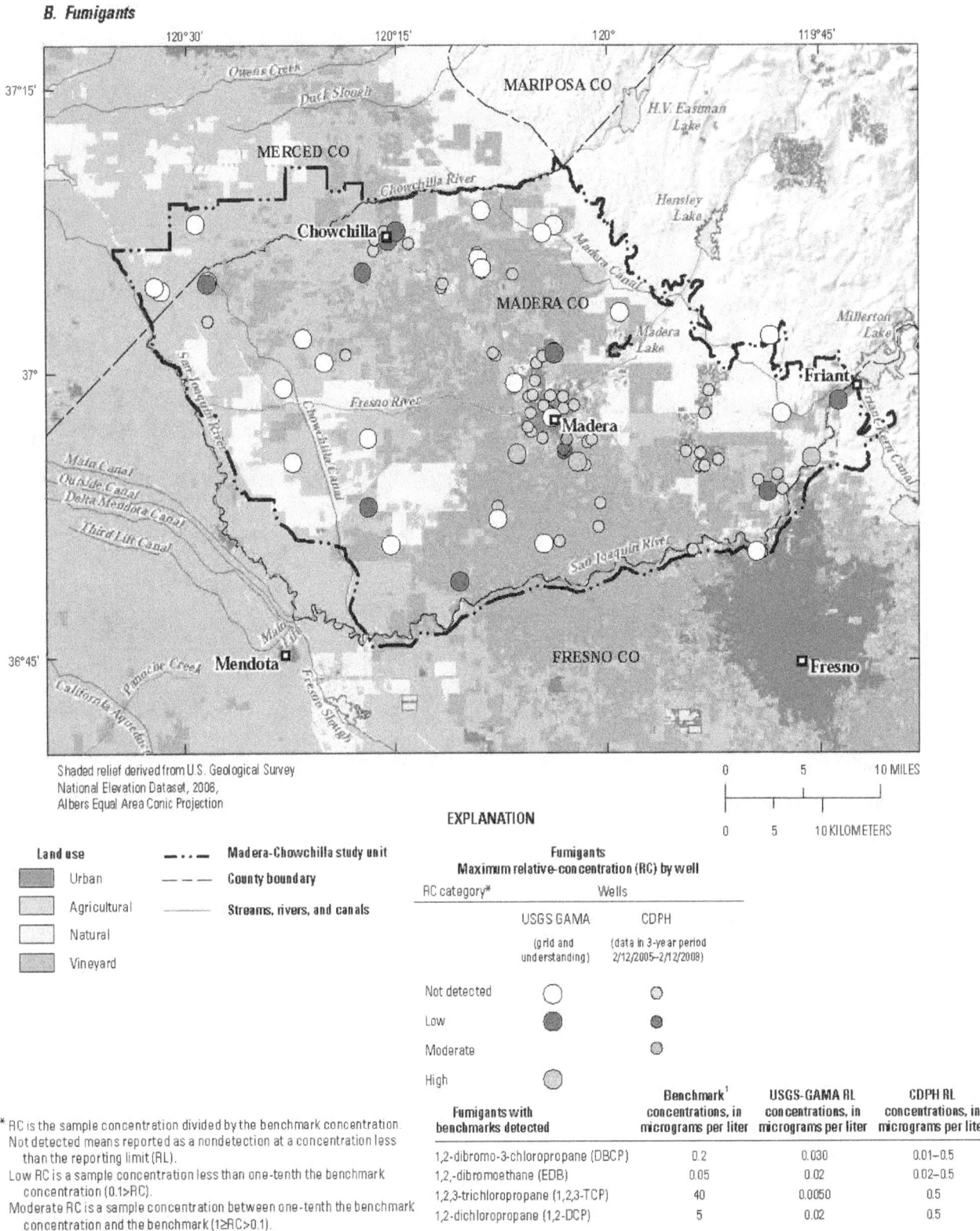

Shaded relief derived from U.S. Geological Survey
National Elevation Dataset, 2008,
Albers Equal Area Conic Projection

EXPLANATION

Land use		Madera-Chowchilla study unit
Urban	— ··· —	
Agricultural	— — —	County boundary
Natural	———	Streams, rivers, and canals
Vineyard		

Fumigants
Maximum relative-concentration (RC) by well

RC category*	Wells	
	USGS GAMA	CDPH
	(grid and understanding)	(data in 3-year period 2/12/2005–2/12/2009)
Not detected	○	○
Low	●	●
Moderate		●
High	○	

* RC is the sample concentration divided by the benchmark concentration.
Not detected means reported as a nondetection at a concentration less
  than the reporting limit (RL).
Low RC is a sample concentration less than one-tenth the benchmark
  concentration (0.1>RC).
Moderate RC is a sample concentration between one-tenth the benchmark
  concentration and the benchmark (1≥RC>0.1).
High RC is a sample concentration greater than the benchmark
  concentration (RC>1).

Fumigants with benchmarks detected	Benchmark[1] concentrations, in micrograms per liter	USGS-GAMA RL concentrations, in micrograms per liter	CDPH RL concentrations, in micrograms per liter
1,2-dibromo-3-chloropropane (DBCP)	0.2	0.030	0.01–0.5
1,2,-dibromoethane (EDB)	0.05	0.02	0.02–0.5
1,2,3-trichloropropane (1,2,3-TCP)	40	0.0050	0.5
1,2-dichloropropane (1,2-DCP)	5	0.02	0.5

[1] See tables 3 and 4 for benchmark types.

Figure 27.—Continued

C. Trihalomethanes (THMs)

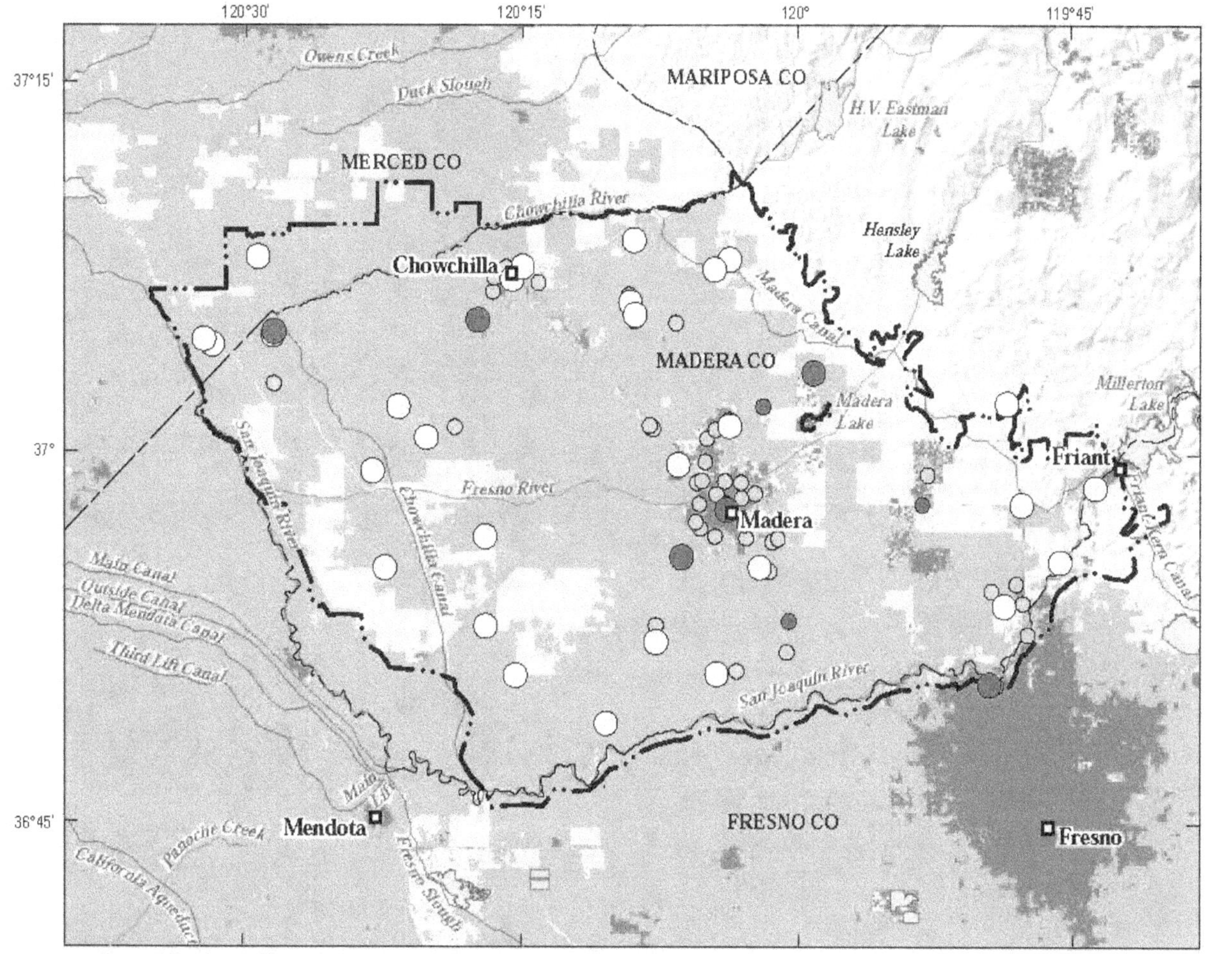

Shaded relief derived from U.S. Geological Survey
National Elevation Dataset, 2006,
Albers Equal Area Conic Projection

EXPLANATION

Land use

Urban	
Agricultural	
Natural	

— · · — Madera-Chowchilla study unit

— — — County boundary

········· Streams, rivers, and canals

Trihalomethane (THM)
Maximum relative-concentration (RC) by well

RC category*	Wells	
	USGS GAMA	CDPH
	(grid and understanding)	(data in 3-year period 2/12/2005–2/12/2008)
Not detected	○	○
Low	●	●

*RC is the sample concentration divided by the benchmark concentration.
Not detected means reported as a nondetection at a concentration less
than the reporting limit (RL).
Low RC is a sample concentration less than one-tenth the benchmark
concentration (0.1>RC).

THMs with benchmarks detected	Benchmark[1] concentrations, in micrograms per liter	USGS-GAMA RL concentrations, in micrograms per liter	CDPH RL concentrations, in micrograms per liter
chloroform	80	0.02	0.5
bromoform	80	0.08	0.5
bromodichloromethane	80	0.04	0.5
dibromochloromethane	80	0.12	0.5

[1] See tables 3 and 4 for benchmark types. The benchmark for THMs is for the sum of
chloroform, bromoform, bromodichloromethane, and dibromochloromethane.

Figure 27.—Continued

D. Solvents

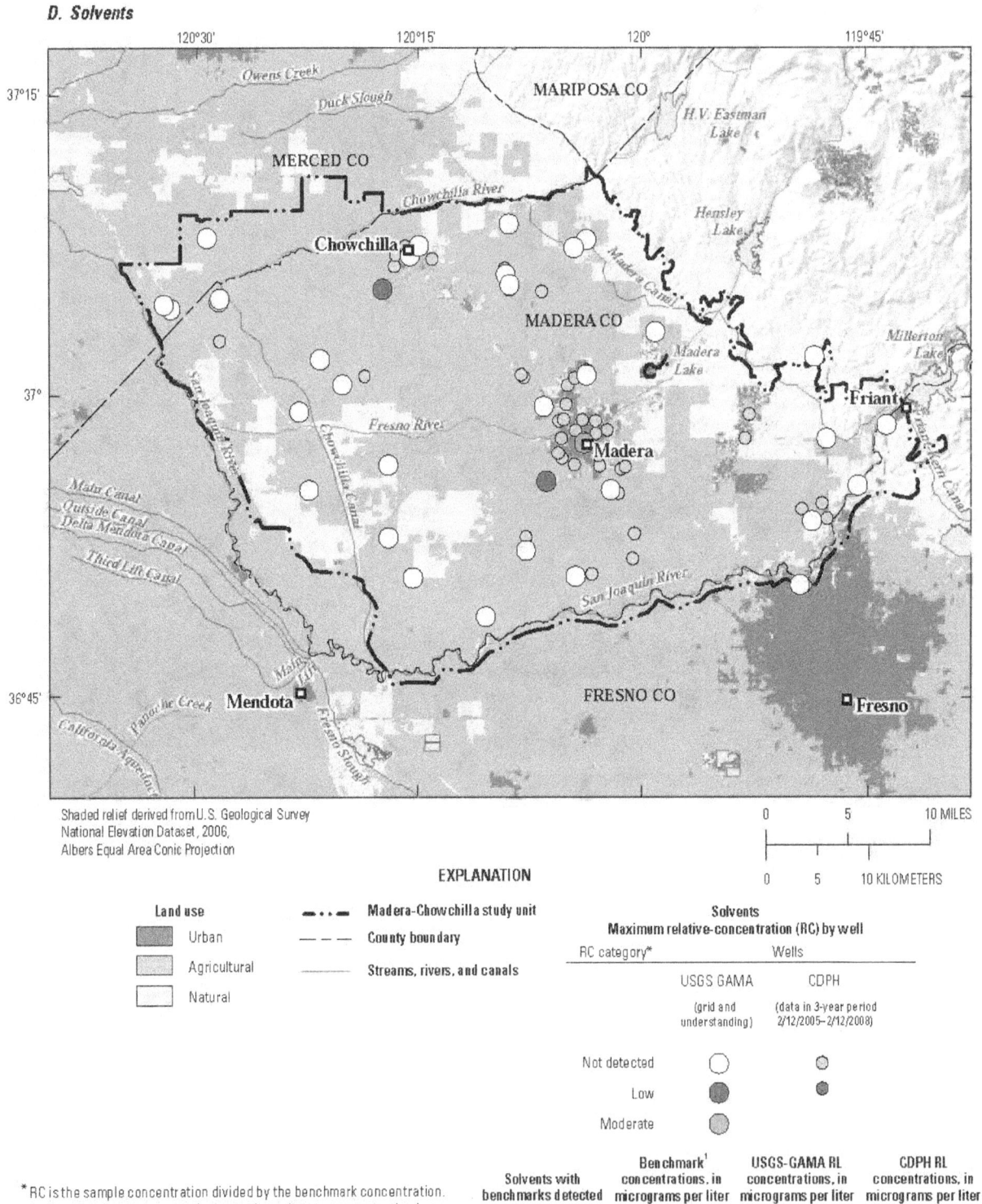

Shaded relief derived from U.S. Geological Survey
National Elevation Dataset, 2006,
Albers Equal Area Conic Projection

EXPLANATION

Land use

Urban

Agricultural

Natural

— · · — Madera-Chowchilla study unit

— — — County boundary

———— Streams, rivers, and canals

Solvents
Maximum relative-concentration (RC) by well

RC category*	Wells	
	USGS GAMA	CDPH
	(grid and understand ing)	(data in 3-year period 2/12/2005–2/12/2008)
Not detected	○	◎
Low	●	◉
Moderate	◐	

*RC is the sample concentration divided by the benchmark concentration.
Not detected means reported as a nondetection at a concentration less
 than the reporting limit (RL).
Low RC is a sample concentration less than one-tenth the benchmark
 concentration (0.1>RC).
Moderate RC is a sample concentration between one-tenth the
 benchmark concentration and the benchmark (1≥RC>0.1).

Solvents with benchmarks detected	Benchmark[1] concentrations, in micrograms per liter	USGS-GAMA RL concentrations, in micrograms per liter	CDPH RL concentrations, in micrograms per liter
tetrachloroethene (PCE)	5	0.04	0.5
trichloroethene (TCE)	5	0.02	0.5
1,1-dichloroethane (1,1-DCA)	5	0.04	0.5
cis-1,2-dichloroethene (cis-1,2-DCE)	6	0.02	not available

[1] See tables 3 and 4 for benchmark types.

Figure 27.—Continued

E. Perchlorate

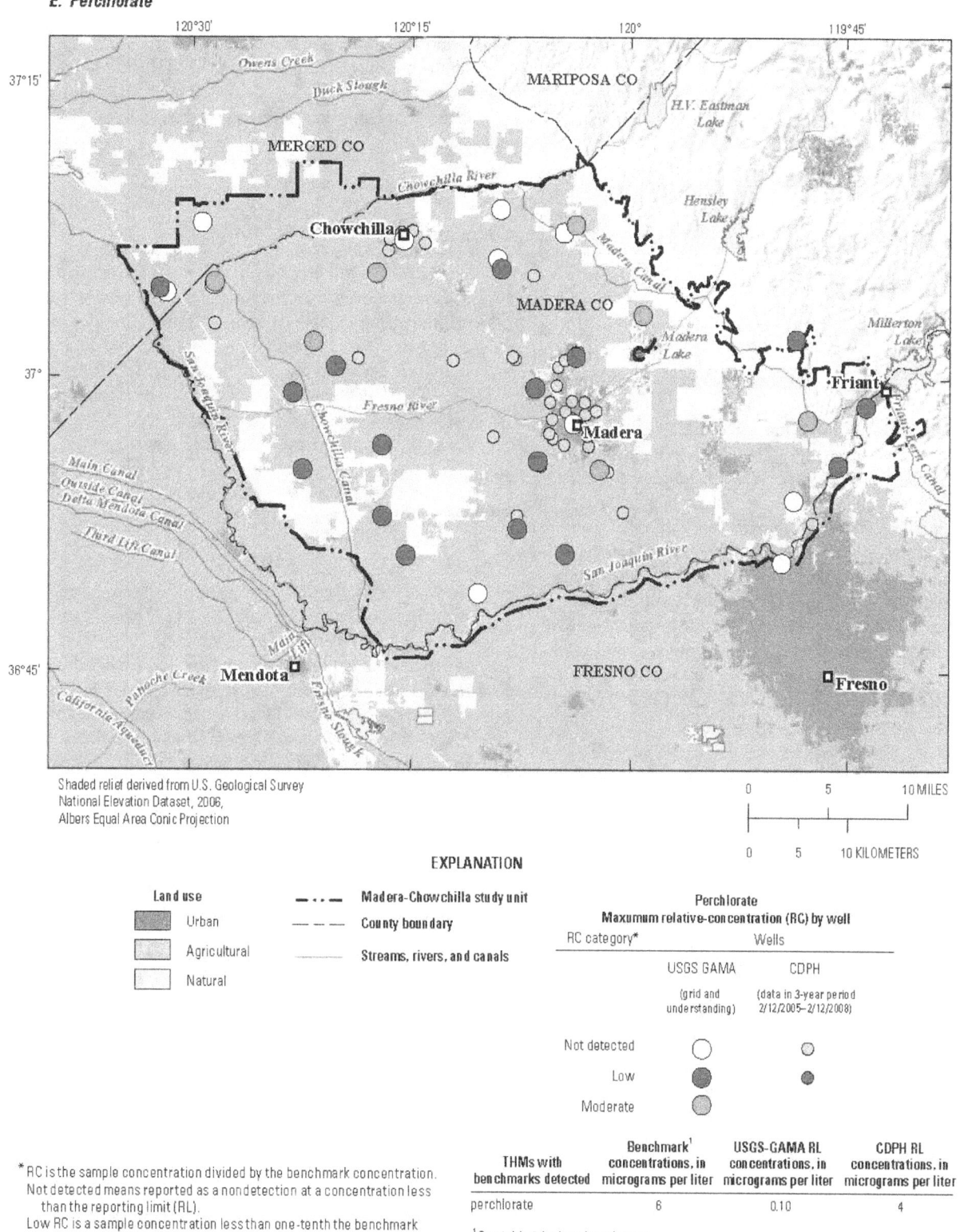

EXPLANATION

Land use
- Urban
- Agricultural
- Natural

— · · · — Madera-Chowchilla study unit
— — — County boundary
——— Streams, rivers, and canals

Perchlorate
Maximum relative-concentration (RC) by well

RC category*	Wells	
	USGS GAMA	CDPH
	(grid and understanding)	(data in 3-year period 2/12/2005–2/12/2008)
Not detected	○	○
Low	●	●
Moderate	◐	

*RC is the sample concentration divided by the benchmark concentration.
 Not detected means reported as a nondetection at a concentration less
   than the reporting limit (RL).
 Low RC is a sample concentration less than one-tenth the benchmark
   concentration (0.1>RC).
 Moderate RC is a sample concentration between one-tenth the
   benchmark concentration and the benchmark (1≥RC>0.1).

THMs with benchmarks detected	Benchmark[1] concentrations, in micrograms per liter	USGS-GAMA RL concentrations, in micrograms per liter	CDPH RL concentrations, in micrograms per liter
perchlorate	6	0.10	4

[1] See table 3 for benchmark types.

Figure 27.—Continued

Of these 240 samples, 24% had detections of DBCP, and 34% had detections of EDB; of the total 139 detections of the two fumigants, 62 had high RCs, and 72 had moderate RCs. Highest concentrations of DBCP and EDB were detected in samples collected at the beginning of the pumping season (approximately June each year), and concentrations decreased through the season (through approximately September each year).

The other fumigants detected by USGS-GAMA, 1,2,3-TCP and 1,2-dichloropropane, were detected at low RCs. Wells with detections are distributed across the study unit (fig. 27B).

Factors Affecting Fumigants

Rather than consider DBCP and 1,2,3-TCP separately in the *understanding assessment*, concentrations of all fumigants were summed and treated as a constituent class. Fumigants were not significantly correlated with any potential explanatory factors (tables 10A,B). Considered independently, DBCP and 1,2,3-TCP also were not significantly correlated with any potential explanatory factors. Fumigants also were not significantly correlated with any of the other water-quality constituents selected for additional evaluation in the *status assessment*. The absence of significant correlations was unexpected, given the results obtained in other studies. Landon and others (2010) reported a significant positive correlation between DBCP and percentage of orchard/vineyard land use. Their findings are consistent with historical use of DBCP on orchards and vineyards in the Central Eastside area of the San Joaquin Valley, which is just to the north of the Madera-Chowchilla study unit. DBCP also was significantly positively correlated with nitrate in the Central Eastside study unit. Burow and others (1999, 2007) investigated occurrence of DBCP in shallow groundwater (60 to 260 ft below land surface) southeast of Fresno. DBCP concentrations generally were highest in groundwater 100 to 230 ft below land surface, where groundwater recharge ages generally corresponded to the time period during which DBCP was used (1955 to 1977), and DBCP had significant positive correlation with nitrate (Burow and others, 1998a, 1998b, 1999, 2007).

The lack of significant relations between fumigant occurrence and land use, groundwater age, depth, or other water-quality constituents in the Madera-Chowchilla study unit may reflect historic usage patterns in the study unit and the depths of wells sampled for this study. DBCP was used between 1955 and 1977, but historic fumigant-use patterns are not well documented. Partial reporting indicates DBCP was used intermittently to treat nematode infestations that primarily occurred in older, well-established vineyards and orchards (Burow and others, 1999); therefore, the use pattern was likely geographically patchy within the study

unit. The fumigant mixture D-D was first used in 1943 (Oki and Giambelluca, 1987); thus, 1,2,3-TCP may have been introduced into groundwater prior to 1950, which may account for the presence of 1,2,3-TCP and 1,2-DCP in pre-modern groundwater in wells with depth to top of perforations greater than 300 ft below land surface (fig. 28B).

Other Volatile Organic Compounds

Water used for drinking water and other household uses in domestic, municipal, and community systems commonly is disinfected with hypochlorite solutions (bleach). In addition to disinfecting the water, the hypochlorite reacts with organic matter to produce trihalomethanes (THMs) and other chlorinated and/or brominated disinfection byproducts. The study-unit detection frequency of the THM chloroform was 17% (fig. 25). No high or moderate RCs of chloroform were detected by USGS-GAMA or reported in the CDPH database (table 8). The maximum RC detected was 0.001 (fig. 26). Most detections of THMs in USGS-GAMA samples or reported in the CDPH database were in wells located in or near the cities of Chowchilla, Madera, and Fresno (fig. 27C).

The sum of THMs was significantly associated with modern groundwater and positively correlated with urban land use (tables 10A,B). Nationally, THMs have also been strongly correlated with urban land use (Zogorski and others, 2006). Potential urban sources of THMs include recharge from landscape irrigation with disinfected water, leakage from water distribution systems, and industrial and commercial usage of chlorinated disinfectants and reagents (Ivahnenko and Barbash, 2004). In addition, shock chlorination is a recommended procedure for treatment of bacterial contamination and odor problems in domestic wells, and may result in a reservoir of chlorinated water in the well bore and surrounding aquifer material (Seiler, 2006). Small systems, such as schools, campgrounds, restaurants, and small community associations, may be likely to maintain their wells following guidelines for domestic wells.

Solvents are used for a variety of industrial, commercial, and domestic purposes (Zogorski and others, 2006). The only frequently detected solvent was PCE, with a study-unit detection frequency of 10% (figs. 25, 26). PCE was also the most frequently detected solvent in groundwater nationally (Zogorski and others, 2006). PCE was present at moderate RCs in 3.3% of the primary aquifer system (table 8). PCE is typically used for dry-cleaning of fabrics and degreasing metal parts and is an ingredient in a wide range of products, including paint removers, polishes, printing inks, lubricants, and adhesives (Doherty, 2000). Detections of solvents in USGS-GAMA samples were in wells located in or near the cities of Chowchilla and Madera (fig. 27D).

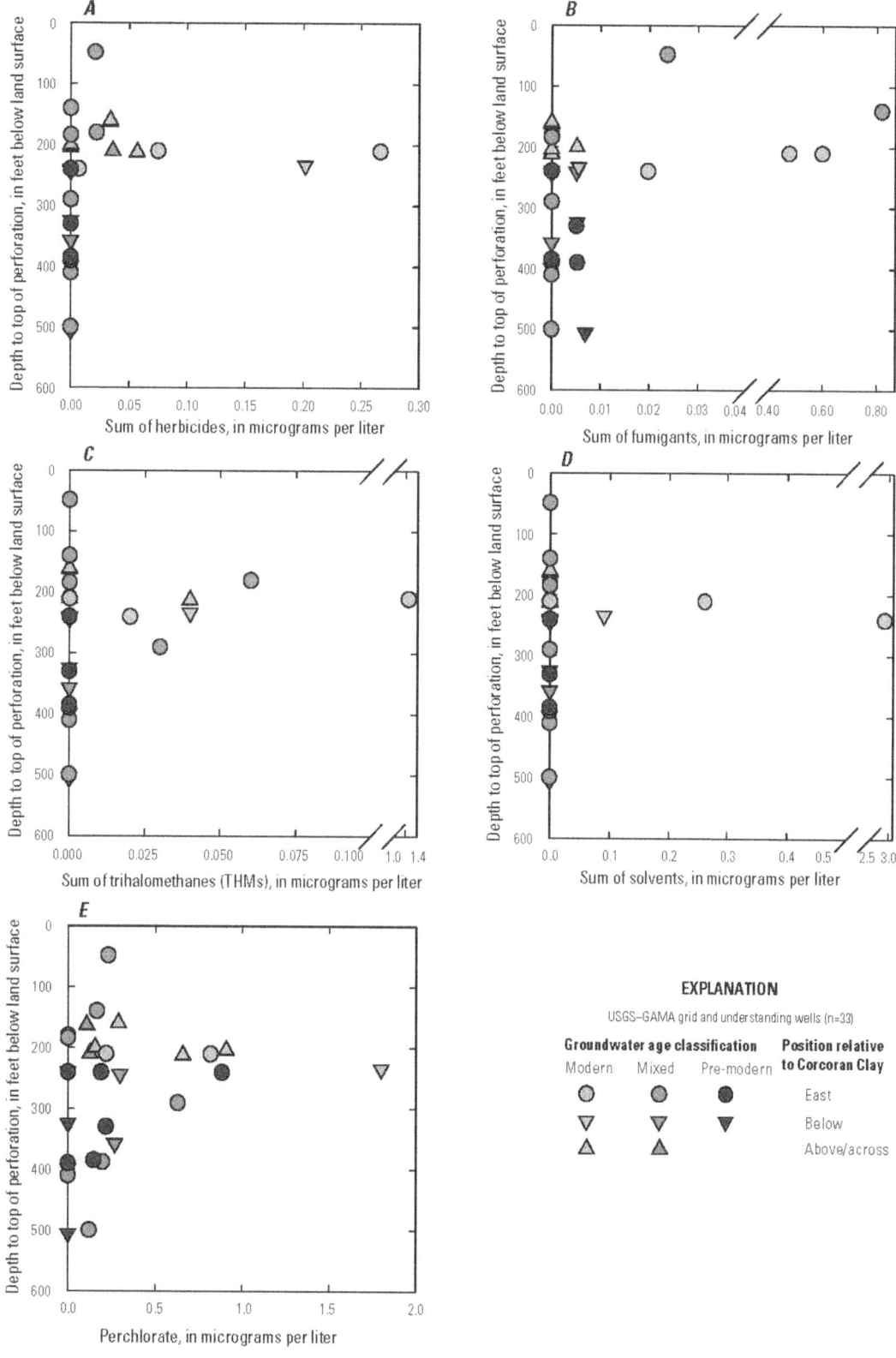

Figure 28.   Relations of (A) herbicide, (B) fumigant, (C) trihalomethane, (D) solvent, and (E) perchlorate total concentrations with depth to top of perforations in wells, Madera-Chowchilla study unit, 2008, California GAMA Priority Basin Project.

Total solvent concentration is the sum of the concentrations of all chlorinated solvents with benchmarks. Three solvents in addition to PCE were detected (table 4), and all detections were in samples that also had detections of PCE (Shelton and others, 2009). The sum of solvents was significantly associated with modern groundwater and positively correlated with urban land use (tables 10A,B).

Special-Interest Constituents

Perchlorate is an inorganic anion with natural and anthropogenic sources. It is formed naturally in the atmosphere and is present in precipitation (Dasgupta and others, 2005; Rajagopalan and others, 2009). Perchlorate salts are the primary ingredient in solid rocket fuel and are used in explosives, safety flares, and fireworks; thus, sites that manufacture, use, or dispose of these products are potential sources of perchlorate contamination to groundwater (U.S. Environmental Protection Agency, 2005). Perchlorate is also present as a contaminant in the Chilean nitrate fertilizer that was used extensively in California before industrial sources of nitrate fertilizers were plentiful (Dasgupta and others, 2006).

Perchlorate was detected at moderate RCs in 20% of the primary aquifer system (table 8) and at low RCs in 50%. No high RCs of perchlorate were detected. Detections of perchlorate were distributed across the study unit (fig. 27E).

Because NDMA was analyzed in only 5 of the 30 grid wells, aquifer-scale proportions for NDMA were not calculated. However, the one detection of NDMA had an RC of 0.25 (moderate).

Factors Affecting Perchlorate

Higher perchlorate concentrations were significantly associated with wells classified as shallow, and perchlorate concentration had a significant negative correlation with well depth (tables 10A,B; fig. 28E). Perchlorate was negatively correlated with pH and positively correlated with dissolved oxygen and Fract-CaMg. The correlations between perchlorate and the geochemical indicators likely reflect the significant associations between high pH and deep wells and between high dissolved oxygen concentrations and Fract-CaMg values and shallow wells.

Fram and Belitz (2011b) investigated the occurrence patterns and concentrations of perchlorate under natural conditions in California. They determined the expected detection frequencies of perchlorate at concentrations greater than threshold concentrations (0.1 µg/L and 0.5 µg/L) under natural conditions as a function of climate. Climate is represented by the aridity index, which is the ratio of average annual precipitation to average annual evapotranspiration (United Nations Educational, Scientific, and Cultural Organization, 1979). For the average aridity

index in the Madera-Chowchilla study unit, the predicted detection frequencies under natural conditions are 63% for concentrations of perchlorate >0.1 µg/L and 12% for concentrations >0.5 µg/L. Perchlorate was detected at a concentration of 0.1 µg/L or higher in 21 of the 30 grid well samples (70%) and it was detected at a concentration of 0.5 µg/L or higher in 7 grid well samples (23%) (Shelton and others, 2009). These observed detection frequencies are greater than the predicted detection frequencies of 63% and 12%, respectively, suggesting that anthropogenic sources or processes have increased the concentrations of perchlorate over natural, background levels in the Madera-Chowchilla study unit. There are no sites of known groundwater contamination from aerospace, military, or industrial sources of perchlorate in the Madera-Chowchilla study unit (U.S. Environmental Protection Agency, 2005; California Department of Toxic Substances Control, 2007; California State Water Resources Control Board, 2007), and the historic use of Chilean nitrate fertilizer is insufficient to account for the excess perchlorate (Dasgupta and others, 2006; Rajagopalan and others, 2006). The inverse correlation between perchlorate and well depth, and the association between higher concentrations of perchlorate and higher concentrations of herbicides, nitrate, and uranium are consistent with the source of the excess perchlorate being remobilization of naturally deposited perchlorate salts in the unsaturated zone by irrigation recharge (Fram and Belitz, 2011b).

Summary

Groundwater quality in the approximately 860-square-mile Madera and Chowchilla Subbasins (Madera-Chowchilla study unit) of the San Joaquin Valley Basin was investigated as part of the Priority Basin Project of the Groundwater Ambient Monitoring and Assessment (GAMA) Program. The study unit is located in California's Central Valley region in parts of Madera, Merced, and Fresno Counties. The GAMA Priority Basin Project is being conducted by the California State Water Resources Control Board in collaboration with the U.S. Geological Survey (USGS) and the Lawrence Livermore National Laboratory. The Priority Basin Project was designed to provide statistically robust assessments of untreated groundwater quality within the primary aquifer systems in California. The primary aquifer system within each study unit is defined by the depth of the perforated or open intervals of the wells listed in the California Department of Public Health (CDPH) database of wells used for public drinking-water supply. The quality of groundwater in shallower or deeper water-bearing zones may differ from that in the primary aquifer system; shallower groundwater may be more vulnerable to contamination from the land surface.

The assessments for the Madera-Chowchilla study unit were based on water-quality and ancillary data collected by the USGS from 35 wells during April–May 2008, and on water-quality data reported in the CDPH database for 125 wells between February 2005 and February 2008. Two types of assessments were made: *status*, which is an assessment of the current quality of the groundwater resource; and *understanding*, which includes identification of natural factors and human activities that may be affecting groundwater quality.

Relative-concentrations (sample concentrations divided by benchmark concentrations) were used for evaluating groundwater quality for those constituents that have Federal and (or) California regulatory or non-regulatory benchmarks for drinking-water quality. A relative-concentration (RC) greater than (>) 1.0 indicates a concentration above a benchmark. RCs for organic constituents (volatile organic compounds and pesticides) and special-interest constituents (perchlorate) were classified as "high" (RC>1.0), "moderate" (1.0≥RC>0.1), or "low" (RC≤0.1). For inorganic constituents (major ions, trace elements, nutrients, and radioactive constituents), the boundary between low and moderate RCs was set at 0.5. The assessments characterize untreated groundwater quality, not the quality of treated drinking water delivered to consumers by water purveyors; drinking-water benchmarks and thus, RCs, are used to provide context for the concentrations of constituents measured in groundwater.

Aquifer-scale proportion was used in the *status assessment* as the primary metric for evaluating regional-scale groundwater quality. High aquifer-scale proportion is defined as the percentage of the area of the primary aquifer system with an RC greater than 1.0 for a particular constituent or class of constituents; moderate and low aquifer-scale proportions are defined as the percentages of the area of the primary aquifer system with moderate and low RCs, respectively. Percentages are based on an areal basis, rather than on a volumetric basis. Two statistical approaches—grid-based, which used one value per grid cell, and spatially weighted, which used multiple values per grid cell—were used to calculate aquifer-scale proportions for individual constituents and classes of constituents. The spatially weighted estimates of high aquifer-scale proportions were within the 90 percent (%) confidence intervals of the grid-based estimates for all constituents except iron.

The *status assessment* indicated that inorganic constituents had greater high and moderate aquifer-scale proportions in the Madera-Chowchilla study unit than did organic constituents. RCs for inorganic constituents with health-based benchmarks were high in 37% of the primary aquifer system, moderate in 30%, and low in 33%. The inorganic constituents contributing most to the high aquifer-scale proportion were arsenic (13%), uranium (17%), gross

alpha particle activity (20%), nitrate (6.7%), and vanadium (3.3%). RCs for inorganic constituents with secondary maximum contaminant levels (non-health-based benchmarks) were high in 6.7% of the primary aquifer system, moderate in 17%, and low in 77%. The constituent contributing most to the high aquifer-scale proportion was total dissolved solids (TDS) (6.7%). RCs for organic constituents with health-based benchmarks were high in 10% of the primary aquifer system, moderate in 3.3%, and low in 40%; organic constituents were not detected in 47% of the primary aquifer system. The fumigant 1,2-dibromo-3-chloropropane (DBCP) was the only organic constituent detected at high RCs. Seven organic constituents were detected in 10% or more of the primary aquifer system: DBCP; the fumigant additive 1,2,3-trichloropropane (1,2,3-TCP); the herbicides simazine, atrazine, and diuron; the trihalomethane (THM) chloroform; and the solvent tetrachloroethene (PCE). RCs for the special-interest constituent perchlorate were moderate in 20% of the primary aquifer system.

The second component of this study, the *understanding assessment*, identified the natural and human factors that may affect groundwater quality by evaluating statistical correlations between water-quality constituents and potential explanatory factors, such as land use, position relative to important geologic features, groundwater age, well depth, and geochemical conditions in the aquifer. Results of the statistical evaluations were used to explain the distribution of constituents in the study unit. Depth to the top of perforations in the well and groundwater age were the most important explanatory factors for many constituents. High and moderate RCs of nitrate, uranium, and TDS, and the presence of herbicides, trihalomethanes, and solvents were associated with depths to the top of perforations less than 235 ft and with modern- and mixed-age groundwater. Positive correlations between uranium, bicarbonate, TDS, and proportion of calcium and magnesium in the total cations suggest that downward movement of recharge from irrigation water contributed to the elevated concentrations of these constituents in the primary aquifer system. High and moderate RCs of arsenic were associated with depths to the top of perforations greater than 235 ft, mixed- and pre-modern-age groundwater, and location in sediments from the Chowchilla River alluvial fan, suggesting that increased residence time and appropriate aquifer materials were needed for arsenic to accumulate in the groundwater. High and moderate RCs of fumigants were associated with depths to the top of perforations of less than 235 ft and location south of the city of Madera; low RCs of fumigants were detected in wells dispersed across the study unit with a range of depths to top of perforations. Land use generally was not a significant explanatory factor, likely because more than 50% of the area within 500 meters of two-thirds of the grid wells was classified as agricultural.

Acknowledgments

The authors thank the following cooperators for their support: the State Water Resources Control Board (SWRCB), California Department of Public Health, California Department of Water Resources, and Lawrence Livermore National Laboratory. We especially thank the cooperating well owners and water purveyors for their generosity in allowing the U.S. Geological Survey (USGS) to collect samples from their wells. Funding for this work was provided by State of California bonds authorized by Proposition 50 and administered by the SWRCB. We also thank the USGS-GAMA team for the many tasks required to complete this work.

References Cited

Aeschbach-Hertig, W., Peeters, F., Beyerle, U., and Kipfer, R., 1999, Interpretation of dissolved atmospheric noble gases in natural waters: Water Resources Research, v. 35, no. 9, p. 2779–2792.

Aeschbach-Hertig, W., Peeters, F., Beyerle, U., and Kipfer, R., 2000, Paleotemperature reconstruction from noble gases in ground water taking into account equilibration with entrapped air: Nature, v. 405, p. 1040–1044.

Anderson, J.R., Handy, E.E., Roach, J.T., and Witmer, R.E., 1976, A land use and land cover classification system for use with remote sensor data: U.S. Geological Survey Professional Paper 964, 28 p., also available at http://pubs.er.usgs.gov/usgspubs/pp/pp964.

Andrews, J.N., 1985, The isotopic composition of radiogenic helium and its use to study groundwater movement in confined aquifers: Chemical Geology, v. 49, p. 339–351.

Andrews, J.N., and Lee, D.J., 1979, Inert gases in groundwater from the Bunter Sandstone of England as indicators of age and paleoclimatic trends: Journal of Hydrology, v. 41, p. 233–252.

Appelo, C.A.J., and Postma, D., 2005, Geochemistry, groundwater, and pollution (2d ed.): Leiden, Germany, A.A. Balkema Publishers, 649 p.

Arndt, M.F., 2010, Evaluation of gross alpha and uranium measurements for MCL compliance: Water Research Foundation, 299 p.

Belitz, K., Dubrovsky, N.M., Burow, K.R., Jurgens, B.C., and Johnson, T., 2003, Framework for a ground-water quality monitoring and assessment program for California: U.S. Geological Survey Water-Resources Investigations Report 03-4166, 78 p.

Belitz, K., Jurgens, B., Landon, M.K., Fram, M.S., and Johnson, T., 2010, Estimation of aquifer-scale proportion using equal-area grids—Assessment of regional-scale groundwater quality: Water Resources Research, v. 46, W11550.

Bennett, G.L., V, Fram, M.S., Belitz, Kenneth, and Jurgens, B.C., 2010, Status and understanding of groundwater quality in the northern San Joaquin Basin, 2005—California GAMA Priority Basin Project: U.S. Geological Survey Scientific Investigations Report 2010–5175, 82 p.

Bertoldi, G.L., Johnston, R.H., and Evenson, K.D., 1991, Ground water in the Central Valley, California—A Summary Report: U.S. Geological Survey Professional Paper 1401-A, 44 p.

Brown, L.D., Cai, T.T., and Dasgupta, A., 2001, Interval estimation for a binomial proportion: Statistical Science, v. 16, no. 2, p. 101–117.

Burow, K.R., Shelton, J.L., and Dubrovsky, N.M., 1998a, Occurrence of nitrate and pesticides in ground water beneath three agricultural land-use settings in the eastern San Joaquin Valley, California: U.S. Geological Survey Water-Resources Investigations Report 97-4284, 51 p.

Burow, K.R., Stork, S.V., and Dubrovsky, N.M., 1998b, Nitrate and pesticides in ground water in the eastern San Joaquin Valley, California—Occurrence and trends: U.S. Geological Survey Water-Resources Investigations Report 98-4040A, 33 p.

Burow, K.R., Panshin, S.Y., Dubrovsky, N.M., Vanbrocklin, D., and Fogg, G.E., 1999, Evaluation of processes affecting 1,2-dibromo-3-chloropropane (DBCP) concentrations in ground water in the Eastern San Joaquin Valley, California—Analysis of chemical data and ground-water flow and transport simulations: U.S. Geological Survey Water-Resources Investigations Report 99-4059, 57 p.

Burow, K.R., Shelton, J.L., Hevesi, J.A., and Weissmann, G.S., 2004, Hydrogeologic characterization of the Modesto area, San Joaquin Valley, California: U.S. Geological Survey Scientific Investigations Report 2004-5232, 54 p.

Burow, K.R., Dubrovsky, N.M., and Shelton, J.L., 2007, Temporal trends in concentrations of DBCP and nitrate in ground water in the eastern San Joaquin Valley, California, USA: Hydrogeology Journal, v. 15, no. 5, p. 991–1007.

Burow, K.R., Shelton, J.L., and Dubrovsky, N.M., 2008, Regional nitrate and pesticide trends in ground water in the eastern San Joaquin Valley, California: Journal of Environmental Quality, v. 37, no. 5, p. S-249–S-263.

California Department of Public Health, 2008a, California drinking water-related laws—Drinking water-related regulations (Title 22), accessed February 5, 2008, at http://www.cdph.ca.gov/certlic/drinkingwater/Pages/Lawbook.aspx.

California Department of Public Health, 2008b, Drinking water notification levels—Notification levels, accessed February 5, 2008, at http://www.cdph.ca.gov/certlic/drinkingwater/Pages/NotificationLevels.aspx.

California Department of Public Health, 2009, 1,2,3-Trichloropropane, accessed November 20, 2009, at http://www.cdph.ca.gov/certlic/drinkingwater/Pages/123TCP.aspx.

California Department of Toxic Substances Control, 2007, EnviroStor: Data download, accessed November 2007 at http://www.envirostor.dtsc.ca.gov/public/data_download.asp.

California Department of Water Resources, 1966, Madera area investigation: California Department of Water Resources Bulletin No. 135, 226 p., 23 pls.

California Department of Water Resources, 2000, Land use data, available at http://www.water.ca.gov/landwateruse/lusrvymain.cfm.

California Department of Water Resources, 2003, California's groundwater: California Department of Water Resources Bulletin 118, 246 p., available at http://www.water.ca.gov/groundwater/bulletin118/update2003.cfm.

California Department of Water Resources, 2004a, California's groundwater—Individual basin descriptions, Chowchilla Subbasin: California Department of Water Resources Bulletin 118, available at http://www.water.ca.gov/groundwater/bulletin118/update2003.cfm.

California Department of Water Resources, 2004b, California's groundwater—Individual basin descriptions, Madera Subbasin: California Department of Water Resources Bulletin 118, available at http://www.water.ca.gov/groundwater/bulletin118/update2003.cfm.

California State Water Resources Control Board, 2002, DBCP Groundwater Information Sheet, available at http://www.swrcb.ca.gov/water_issues/programs/gama/docs/coc_dbcb_infosheet_jz0610.pdf.

California State Water Resources Control Board, 2003, A comprehensive groundwater quality monitoring program for California: Assembly Bill 599 Report to the Governor and Legislature, March 2003, 100 p., accessed February 6, 2012, at http://www.waterboards.ca.gov/gama/docs/final_ab_599_rpt_to_legis_7_31_03.pdf.

California State Water Resources Control Board, 2007, GeoTracker—Cleanup sites download, accessed November 2007 at http://www.geotracker.waterboards.ca.gov/.

Chapelle, F.H., 2001, Ground-water microbiology and geochemistry (2d ed.): New York, John Wiley and Sons, Inc., 477 p.

Chapelle, F.H., McMahon, P.B., Dubrovsky, N.M., Fuji, R.F., Oaksford, E.T., and Vroblesky, D.A., 1995, Deducing the distribution of terminal electron-accepting processes in hydrologically diverse groundwater systems: Water Resources Research, v. 31, no. 2, p. 359–371.

Clark, I.D., and Fritz, P., 1997, Environmental isotopes in hydrogeology: New York, Lewis Publishers, 328 p.

Cook, P.G., and Böhlke, J.K., 2000, Determining timescales for groundwater flow and solute transport, in Cook, P.G., and Herczeg, A., eds., Environmental tracers in subsurface hydrology: Boston, Kluwer Academic Publishers, p. 1–30.

Craig, H., and Lal, D., 1961, The production rate of natural tritium: Tellus, v. 13, p. 85–105.

Dasgupta, P.K., Dyke, J.V., Kirk, A.B., and Jackson, W.A., 2006, Perchlorate in the United States—Analysis of relative source contributions to the food chain: Environmental Science and Technology, v. 40, p. 6608–6614.

Dasgupta, P.K., Martinelango, P.K., Jackson, W.A., Anderston, T.A., Tian, K., Tock, R.W., and Rajagopalan, S., 2005, The origin of naturally occurring perchlorate—The role of atmospheric processes: Environmental Science and Technology, v. 39, p. 1569–1575.

Davis, S., and DeWiest, R.J., 1966, Hydrogeology: New York, John Wiley and Sons, 413 p.

Davis, S.N., and Hall, F.R., 1959, Water-quality of eastern Stanislaus and northern Merced counties, California: Stanford University Publications, Geological Sciences, v. 6, no. 1, 112 p.

Doherty, R.E., 2000, A history of the production and use of carbon tetrachloride, tetrachloroethylene, trichloroethylene, and 1,1,1-trichloroethane in the United States—Part 1. Historical background; carbon tetrachloride and tetrachloroethylene: Journal of Environmental Forensics, v. 1, p. 69–81.

Domagalski, J.L., 1997, Pesticides in surface and ground water of the San Joaquin-Tulare basins, California—Analysis of available data, 1966 through 1992: U.S. Geological Survey Water-Supply Paper 2468, 74 p.

Dubrovsky, N.M., Kratzer, C.R., Brown, L.R., Gronberg, J.M., and Burow, K.R., 1998, Water quality in the San Joaquin–Tulare Basins, California, 1992–95: U.S. Geological Survey Circular 1159, 38 p.

Faunt, C.C., ed., 2009, Groundwater availability of the Central Valley Aquifer, California: U.S. Geological Survey Professional Paper 1766, 225 p.

Fontes, J.C., and Garnier, J.M., 1979, Determination of the initial ^{14}C activity of the total dissolved carbon—A review of the existing models and a new approach: Water Resources Research, v. 15, p. 399–413.

Fram, M.S., and Belitz, Kenneth, 2011a, Occurrence and concentrations of pharmaceutical compounds in groundwater used for public drinking-water supply in California: Science of the Total Environment, v. 409, no. 18, p. 3409–3417, accessed February 6, 2012, at http://www.sciencedirect.com/science/article/pii/S0048969711005778.

Fram, M.S., and Belitz, K., 2011b, Probability of detecting perchlorate under natural conditions in deep groundwater in California and the Southwestern United States: Environmental Science & Technology, v. 45, no. 4, p. 1271–1277.

Gilliom, R.J., Barbash, J.E., Crawford, C.G., Hamilton, P.A., Martin, J.D., Nakagaki, N., Nowell, L.H., Scott, J.C., Stackelberg, P.E., Thelin, G.P., and Wolock, D.M., 2006, The quality of our nation's waters—Pesticides in the nation's streams and ground water, 1992–2001: U.S. Geological Survey Circular 1291, 172 p.

Gronberg, J.M., Dubrovsky, N.M., Kratzer, C.R., Domagalski, J.L., Brown, L.R., and Burow, K.R., 1998, Environmental setting and study design for assessing water quality in the San Joaquin–Tulare Basins, California: U.S. Geological Survey Water-Resources Investigations Report 97–4205, 45 p.

Helsel, D.R., and Hirsch, R.M., 2002, Statistical methods in Water Resources: U.S. Geological Survey Techniques of Water-Resources Investigations, book 4, chap. A3, 510 p., available at http://water.usgs.gov/pubs/twri/twri4a3/.

Hem, J.D., 1992, Study and interpretation of the chemical characteristics of natural water (3d ed.): U.S. Geological Survey Water-Supply Paper 2254, 263 p.

Huntington, G.L., 1971, Soil survey, eastern Fresno area, California: U.S. Department of Agriculture, Soil Conservation Service.

Isaaks, E.H., and Srivastava, R.M., 1989, Applied geostatistics: New York, Oxford University Press, 561 p.

Ivahnenko, Tammy, and Barbash, J.E., 2004, Chloroform in the hydrologic system—Sources, transport, fate, occurrence, and effects on human health and aquatic organisms: U.S. Geological Survey Scientific Investigations Report 2004-5137, 34 p.

Izbicki, J.A., Metzger, L.F., McPherson, K.R., Everett, R.R., and Bennett, G.L., 2006, Sources of high-chloride water to wells, Eastern San Joaquin Ground-Water Subbasin, California: U.S. Geological Survey Open-File Report 2006-1309, 8 p.

Izbicki, J.A., Stamos, C.L., Metzger, L.F., Halford, K.J., Kulp, T.R., and Bennett, G.L., 2008, Source, distribution, and management of arsenic in water from wells, Eastern San Joaquin Ground-Water Subbasin, California: U.S. Geological Survey Open-File Report 2008-1272, 8 p.

Jennings, C.W., 1977, Geologic map of California: California Department of Conservation, Division of Mines and Geology, Geologic Data Map No. 2, scale 1:750,000.

Johnson, T.D., and Belitz, K., 2009, Assigning land use to supply wells for the statistical characterization of regional groundwater quality—Correlating urban land use and VOC occurrence: Journal of Hydrology, v. 370, p. 100–108.

Jurgens, B.C., Burow, K.R., Dalgish, B.A., and Shelton, J.L., 2008, Hydrogeology, water chemistry, and factors affecting the transport of contaminants in the zone of contribution to a public-supply well in Modesto, eastern San Joaquin Valley, California: U.S. Geological Survey Scientific Investigations Report 2008-5156, 78 p., available at http://pubs.usgs.gov/sir/2008/5156/.

Jurgens, B.C., Fram, M.S., Belitz, K., Burow, K.R., and Landon, M.K., 2009a, Effects of groundwater development on uranium—Central Valley, California, USA: Ground Water, v. 48, no. 6, p. 913–928.

Jurgens, B.C., McMahon, P.B., Chapelle, F.H., and Eberts, S.M., 2009b, An Excel workbook for identifying redox processes in groundwater: U.S. Geological Survey Open-File Report 2009-1004, 8 p.

Kegley, S.E., Hill, B.R., Orme, S., and Choi, A.H., 2008, PAN Pesticide Database, Pesticide Action Network, North America: San Francisco, accessed February 6, 2012, at http://www.pesticideinfo.org.

Kenny, J.F., Barber, N.L., Hutson, S.S., Linsey, K.S., Lovelace, J.K., and Maupin, M.A., 2009, Estimated use of water in the United States in 2005: U.S. Geological Survey Circular 1344, 53 p.

Kolpin, D.W., Thurman, E.M., and Linhart, S.M., 1998, The environmental occurrence of herbicides—The importance of degradates in ground water: Archives of Environmental Contamination and Toxicology, v. 35, p. 385–390.

Kulongoski, J., and Belitz, K., 2004, Ground-water ambient monitoring and assessment program: U.S. Geological Survey Fact Sheet 2004-3088.

Kulongoski, J.T., Hilton, D.R., Cresswell, R.G., Hostetler, S., and Jacobson, G., 2008, Helium-4 characteristics of groundwaters from Central Australia—Comparative chronology with chlorine-36 and carbon-14 dating techniques: Journal of Hydrology, v. 348, p. 176–194.

Landon, M.K., Belitz, K., Jurgens, B.C., Kulongoski, J.T., and Johnson, T., 2010, Status and understanding of groundwater quality in the Central-Eastside San Joaquin basin, 2006—California GAMA Priority Basin Project: U.S. Geological Survey Scientific Investigations Report 2009-5266, 97 p.

Lucas, L.L., and Unterweger, M.P., 2000, Comprehensive review and critical evaluation of the half-life of tritium: Journal of Research of the National Institute of Standards and Technology, v. 105, no. 4, p. 541–549.

Madera Irrigation District, 2004, Madera Irrigation District Map, accessed February 6, 2012, at http://madera-id.org/images/pdf/District_Map.pdf.

Marchand, D.E., and Allwardt, Alan, 1981, Late Cenozoic stratigraphic units, northeastern San Joaquin Valley, California: U.S. Geological Survey Bulletin 1470, 70 p.

McMahon, P., Böhlke, J.K., Kauffman, L.J., Kipp, K.L., Landon, M.K., Crandall, C.A., Burow, K.R., and Brown, C.J., 2008, Source and transport controls on the movement of nitrate to public supply wells in selected principal aquifers of the United States: Water Resources Research, v. 44, no. W04401, doi:10.1029/2007WR006252.

McMahon, P.B., and Chapelle, F.H., 2008, Redox processes and water quality of selected principal aquifer systems: Ground Water, v. 46, no. 2, p. 259–271.

Mendenhall, W.C., 1908, Preliminary report on the ground waters of San Joaquin Valley, California: U.S. Geological Survey Water-Supply Paper 222, 52 p.

Mendenhall, W.C., Dole, R.B., and Stabler, H., 1916, Ground water in San Joaquin Valley, California: U.S. Geological Survey Water-Supply Paper 398, 310 p.

Michel, R., and Schroeder, R., 1994, Use of long-term tritium records from the Colorado River to determine timescales for hydrologic processes associated with irrigation in the Imperial Valley, California: Applied Geochemistry, v. 9, p. 387–401.

Michel, R.L., 1989, Tritium deposition in the continental United States, 1953–83: U.S. Geological Survey Water-Resources Investigations Report 89-4072, 46 p.

Mitten, H.T., LeBlanc, R.A., and Bertoldi, G.L., 1970, Geology, hydrology, and quality of water in the Madera area, San Joaquin Valley, California: U.S. Geological Survey Open-File Report.

Morrison, P., and Pine, J., 1955, Radiogenic origin of helium isotopes in rock: Annals of the New York Academy of Sciences, v. 12, p. 19–92.

Nakagaki, N., Price, C.V., Falcone, J.A., Hitt, K.J., and Ruddy, B.C., 2007, Enhanced National Land Cover Data 1992 (NLCDe 92): U.S. Geological Survey Raster digital data, available online at http://water.usgs.gov/lookup/getspatial?nlcde92.

Nakagaki, N., and Wolock, D.M., 2005, Estimation of agricultural pesticide use in drainage basins using land cover maps and county pesticide data: U.S. Geological Survey Open-File Report 05-1188.

Oki, D.S., and Giambelluca, T.W., 1987, DBCP, EDB, and TCP contamination of ground water in Hawaii: Ground Water, v. 25, no. 6, p. 693–702.

Page, R.W., 1986, Geology of the fresh ground-water basin of the Central Valley, California, with texture maps and sections: U.S. Geological Survey Professional Paper 1401-C, 54 p.

Phillips, S.P., Green, C.T., Burow, K.R., Shelton, J.L., and Rewis, D.L., 2007, Simulation of multiscale ground-water flow in part of the northeastern San Joaquin Valley, California: U.S. Geological Survey Scientific Investigations Report 2007-5009, 43 p.

Piper, A.M., 1944, A graphic procedure in the geochemical interpretation of water analyses: American Geophysical Union Transactions, v. 25, p. 914–923.

Plummer, L.N., Michel, R.L., Thurman, E.M., and Glynn, P.D., 1993, Environmental tracers for age-dating young ground water, in Alley, W.M., ed., Regional ground-water quality: New York, Van Nostrand Reinhold, p. 255–294.

Poreda, R.J., Cerling, T.E., and Salomon, D.K., 1988, Tritium and helium isotopes as hydrologic tracers in a shallow unconfined aquifer: Journal of Hydrology, v. 103, p. 1–9.

Rajagopalan, S., Anderson, T.A., Fahlquist, L., Rainwater, K.A., Ridley, M., and Jackson, W.A., 2006, Widespread presence of naturally occurring perchlorate in the high plains of Texas and New Mexico: Environmental Science and Technology, v. 40, p. 3156–3162.

Rajagopalan, S., Anderson, T.A., Cox, S., Harvey, G., Cheng, Q., and Jackson, W.A., 2009, Perchlorate in wet deposition across North America: Environmental Science & Technology, v. 43, no. 3, p. 616–622.

Rowe, B.L., Toccalino, P.L., Moran, M.J., Zogorski, J.S., and Price, C.V., 2007, Occurrence and potential human-health relevance of volatile organic compounds in drinking water from domestic wells in the United States: Environmental Health Perspectives, v. 115, no. 11, p. 1539–1546.

Saucedo, G.J., Bedford, D.R., Raines, G.L., Miller, R.J., and Wentworth, C.M., 2000, GIS data for the geologic map of California (version 2.0): Sacramento, California, California Department of Conservation, Division of Mines and Geology.

Scott, J.C., 1990, Computerized stratified random site selection approaches for design of a ground-water quality sampling network: U.S. Geological Survey Water-Resources Investigations Report 90-4101, 109 p.

Seiler, R.L., 2006, Mobilization of lead and other trace elements following shock chlorination of wells: Science of the Total Environment, v. 367, p. 757–768.

Shelton, J.L., Fram, M.S., and Belitz, Kenneth, 2009, Groundwater-quality data for the Madera–Chowchilla study unit, 2008—Results from the California GAMA Program: U.S. Geological Survey Data Series 455, 80 p.

State of California, 1999, Supplemental report of the 1999 Budget Act 1999–00 Fiscal Year, Item 3940-001-0001, State Water Resources Control Board, accessed August 11, 2010, at http://www.lao.ca.gov/1999/99-00_supp_rpt_lang.html#3940.

State of California, 2001a, Assembly Bill No. 599, Chapter 522, accessed August 11, 2010, at http://www.swrcb.ca.gov/gama/docs/ab_599_bill_20011005_chaptered.pdf.

State of California, 2001b, Groundwater Monitoring Act of 2001: California Water Code, part 2.76, Sections 10780–10782.3, accessed August 11, 2010, at http://www.leginfo.ca.gov/cgibin/displaycode?section=wat&group=10001-11000&file=10780-10782.3.

Stollenwerk, K., 2003, Geochemical processes controlling transport of arsenic in groundwater—A review of adsorption, in Welch, A.H., and Stollenwerk, K.G., eds., Arsenic in ground water—Geochemistry and occurrence: Boston, Kluwer Academic Publishers, 475 p.

Takaoka, N., and Mizutani, Y., 1987, Tritiogenic 3He in groundwater in Takaoka: Earth and Planetary Science Letters, v. 85, p. 74–78.

Toccalino, P.L., 2007, Development and application of health-based screening levels for use in water-quality assessments: U.S. Geological Survey Scientific Investigations Report 2007-5106, 12 p.

Toccalino, P.L., and Norman, J.E., 2006, Health-based screening levels to evaluate U.S. Geological Survey ground water quality data: Risk Analysis, v. 26, no. 5, p. 1339–1348.

Toccalino, P.L., Norman, J.E., Phillips, R.H., Kauffman, L.J., Stackelberg, P.E., Nowell, L.H., Krietzman, S.J., and Post, G.B., 2004, Application of health-based screening levels to ground-water quality data in a state-scale pilot effort: U.S. Geological Survey Scientific Investigations Report 2004-5174, 14 p.

Toccalino, P.L., Norman, J.E., and Hitt, K.J., 2010, Quality of source water from public-supply wells in the United States, 1993–2007: U.S. Geological Survey Scientific Investigations Report 2010-5024, 126 p.

Todd Engineers, 2002, AB3030 Groundwater Management Plan Madera County: Final Draft 2002 accessed October 27, 2008, at http://www.madera-county.com/rma/archives/uploads/1157731120_Document_upload_ab3030plan.pdf.

Tolstikhin, I.N., and Kamensky, I.L., 1969, Determination of ground-water ages by the T-3He method: Geochemistry International, v. 6, p. 810–811.

Torgersen, T., 1980, Controls on pore-fluid concentrations of ^{4}He and ^{222}Rn and the calculation of ^{4}He/^{222}Rn ages: Journal of Geochemical Exploration, v. 13, p. 57–75.

Torgersen, T., and Clarke, W.B., 1985, Helium accumulation in groundwater—I. An evaluation of sources and continental flux of crustal 4He in the Great Artesian basin, Australia: Geochimica et Cosmochimica Acta, v. 49, p. 1211–1218.

Torgersen, T., Clarke, W.B., and Jenkins, W.J., 1979, The tritium/helium-3 method in hydrology: IAEA-SM-228/49, p. 917–930.

Troiano, J., Weaver, D., Marade, J., Spurlock, F., Pepple, M., Nordmark, C., and Bartkowiak, D., 2001, Summary of well water sampling in California to detect pesticide residues resulting from nonpoint source applications: Journal of Environmental Quality, v. 30, p. 448–459.

United Nations Educational, Scientific, and Cultural Organization (UNESCO), 1979, Map of the world distribution of arid regions—Explanatory note: MAB Technical Notes, v. 7, 42 p.

U.S. Census Bureau, 2010, State and county quick facts, California, accessed February 11, 2011, at http://quickfacts.census.gov/qfd/states/06000.html.

U.S. Environmental Protection Agency, 1998, Code of Federal Regulations, title 40—protection of environment, chapter 1—environmental protection agency, subchapter E—pesticide programs, part 159—statements of policies and interpretations, subpart D—reporting requirements for risk/benefit information, 40 CFR 159.184: National Archives and Records Administration, September 19, 1997; amended June 19, 1998, accessed September 5, 2008, at http://www.epa.gov/EPA-PEST/1997/September/Day-19/p24937.htm.

U.S. Environmental Protection Agency, 2005, List of known perchlorate releases in the U.S., March 25, 2005, accessed November 2007 at http://www.epa.gov/sweffrr/documents/perchlorate_releases_us_20050325.htm.

U.S. Environmental Protection Agency, 2008a, Drinking water contaminants, accessed February 5, 2008, at http://www.epa.gov/safewater/contaminants/index.html.

U.S. Environmental Protection Agency, 2008b, Drinking water health advisories—2006 Drinking water standards and health advisory tables, accessed February 5, 2008, at http://www.epa.gov/waterscience/criteria/drinking/.

U.S. Geological Survey, 2010, California Water Science Center—Groundwater Ambient Monitoring and Assessment (GAMA) Program, accessed August 11, 2011, at http://ca.water.usgs.gov/gama/.

Vogel, J.C., and Ehhalt, D., 1963, The use of the carbon isotopes in groundwater studies, in Radioisotopes in hydrology: Tokyo, IAEA, p. 383–395.

Weissmann, G.S., Bennett, G.L., and Lansdale, A.L., 2005, Factors controlling sequence development on Quaternary fluvial fans, San Joaquin Basin, California, USA, in Harvey, A.M., Mather, A.E., and Stokes, M., eds., Alluvial fans—Geomorphology, sedimentology, dynamics: Geological Society of London Special Publication 251, p. 169–186.

Welch, A.H., Westjohn, D.B., Helsel, D.R., and Wanty, R.B., 2000, Arsenic in groundwater of the United States—Occurrence and geochemistry: Ground Water, v. 38, no. 4, p. 589–604.

Welch, A.H., Oremland, R.S., Davis, J.A., and Watkins, S.A., 2006, Arsenic in ground water—A review of current knowledge and relation to the CALFED solution area with recommendations for needed research: San Francisco Estuary and Watershed Science, v. 4, no. 2, article 4, 32 p., accessed May 19, 2008, at http://repositories.cdlib.org/jmie/sfews/vol4/iss2/art4/.

Western Regional Climate Center, 2009, Western Regional Climate Center, Summary climate data for central California, average monthly precipitation data for Madera, California, and for Friant Government Camp, California, accessed May 22, 2009, at http://www.wrcc.dri.edu/summary/Climsmnca.html.

Williamson, A.K., Prudic, D.E., and Swain, L.A., 1989, Groundwater flow in the Central Valley, California: U.S. Geological Survey Professional Paper 1410-D, 127 p.

Wright, M.T., and Belitz, K., 2010, Factors controlling the regional distribution of vanadium in groundwater: Ground Water, v. 48, no. 4, p. 515–525.

Zogorski, J.S., Carter, J.M., Ivahnenko, T., Lapham, W.W., Moran, M.J., Rowe, B.L., Squillace, P.J., and Toccalino, P.L., 2006, Volatile organic compounds in the Nation's ground water and drinking-water supply wells: U.S. Geological Survey Circular 1292, 101 p.

Appendix A. Ancillary Datasets

Land Use

Land use was classified using an "enhanced" version of the satellite-derived (30-m pixel resolution) nationwide USGS National Land Cover Dataset (Nakagaki and others, 2007). This dataset has been used in previous national and regional studies relating land use to water quality (Gilliom and others, 2006; Zogorski and others, 2006). The data represent land use during approximately the early 1990s. The imagery is classified into 25 land-cover classifications (Nakagaki and Wolock, 2005). These 25 land-cover classifications were condensed into 3 principal land-use categories: urban, agricultural, and natural. One subcategory of agricultural land use is orchard/vineyard. Land-use statistics for the study unit, study areas, and for circles with a radius of 500 m around each well were calculated for classified datasets by using ArcGIS (Johnson and Belitz, 2009). A 500-m radius centered on the well has been shown to be effective at correlating urban land use with VOC occurrence for the purposes of statistical characterization (Johnson and Belitz, 2009). Land-use statistics for grid and understanding wells are listed in table A1.

Land-cover classes are based on features distinguishable in Level II remote sensing data (high-altitude aerial photography; Anderson and others, 1976). Urban land use includes high, moderate, and low intensity development and developed open space. Agricultural land use includes cultivated crops and land used for pasture or hay. Natural land use includes everything else. Open-range grazing is classified as natural land use, not agricultural land use.

Lateral Position

The lateral position of wells serves as a proxy for the horizontal position in the regional groundwater flow system. Regionally, groundwater primarily flows from the eastern margins of the valley deposits along the Sierra Mountain front towards the San Joaquin River. The groundwater flow system has vertical flow components as well as horizontal flow components that deviate from the general direction in response to withdrawals and recharge (Phillips and others, 2007). Nevertheless, because the predominant pattern of regional groundwater flow is from the valley margin towards the San Joaquin River, lateral position serves as an approximate indicator of relative position of a well within the regional flow system. The normalized lateral position of each well was calculated as the ratio of the distance from the well to the edge of the regional groundwater flow system to the total distance from the valley trough to the edge of the valley. The eastern edge of the valley was represented by the boundary of the valley fill deposits and was assigned a value of 1.00

(fig. 8). The valley trough was represented by the southeast-to-northwest flowing reach of the San Joaquin River and was assigned a value of 0.00. Both boundaries were represented as approximate line segments, and lateral position was calculated along lines perpendicular to both bounding lines. The normalized lateral position (hereinafter, lateral position) was calculated for all 30- by 30-m-wide cells in the San Joaquin Valley as part of a regional groundwater flow modeling study (Faunt, 2009). Lateral position values were assigned to all wells residing within those cells in ArcGIS (version 9.2) (table A1). Higher values of lateral position indicate locations in the upgradient or proximal portion of the flow system, and lower values of lateral position indicate locations in the downgradient or distal portion of the flow system.

Depth

Well construction data were primarily determined from driller's logs (table A2). In some cases, well construction data were obtained from ancillary records of well owners.

Well construction information, land-surface elevation, and the elevations of the top and bottom of the Corcoran Clay (Page, 1986) were used to code wells as to the depth of the perforated interval relative to the depth and position of the Corcoran Clay (table A2), using ArcGIS information compiled for the Central Valley model (Faunt, 2009). Wells perforated above or across the clay were coded as "Above/Across"; wells perforated below the Corcoran Clay were coded as "Below." Wells located east of the extent of the Corcoran Clay were coded as "East." Well construction information was incomplete for one well, which was coded as "Unknown."

Groundwater Age

Groundwater dating techniques provide a measure of the time since the groundwater was last in contact with the atmosphere. Techniques aimed at estimating groundwater residence times or 'age' include those based on tritium (^{3}H) (for example, Tolstikhin and Kamensky, 1969), carbon-14 (^{14}C) activity (for example, Vogel and Ehhalt, 1963; Plummer and others, 1993), dissolved noble gases, particularly helium-4 (^{4}He) accumulation (for example, Davis and DeWiest, 1966; Andrews and Lee, 1979; Kulongoski and others, 2008), and tritium in combination with its decay product helium-3 (^{3}He) (Poreda and others, 1988; Schlosser and others, 1989).

Tritium (^{3}H) is a short-lived radioactive isotope of hydrogen with a half-life of 12.32 years (Lucas and Unterweger, 2000). Tritium is produced naturally in the atmosphere from the interaction of cosmogenic radiation with nitrogen (Craig and Lal, 1961), by above-ground nuclear weapons testing, and by the operation of nuclear reactors.

Table A1. Land-use information and lateral positions for grid and understanding wells sampled in April and May 2008, Madera-Chowchilla study unit, California GAMA Priority Basin Project.

[USGS-GAMA well identification number: MADCHOW, Madera-Chowchilla study unit grid well; MADCHOWFP, Madera-Chowchilla study unit understanding wells. Other abbreviations: m, meter]

USGS-GAMA well identification number	Land-use information [1]				Normalized lateral position (dimensionless)
	Agricultural land use within 500 m of the well (percent)	Natural land use within 500 m of the well (percent)	Urban land use within 500 m of the well (percent)	Orchard / Vineyard land use within 500 m of the well (percent)	
Grid wells					
MADCHOW-01	45	7	49	40	0.49
MADCHOW-02	9	50	41	6	0.60
MADCHOW-03	0	5	95	0	0.61
MADCHOW-04	12	7	80	0	0.79
MADCHOW-05	68	25	6	66	0.91
MADCHOW-06	12	88	0	11	1.00
MADCHOW-07	57	43	0	0	0.92
MADCHOW-08	99	0	1	2	0.73
MADCHOW-09	59	12	29	14	0.50
MADCHOW-10	0	8	92	0	0.60
MADCHOW-11	95	3	2	38	0.72
MADCHOW-12	64	0	36	0	0.22
MADCHOW-13	99	1	1	99	0.37
MADCHOW-14	97	2	1	97	0.47
MADCHOW-15	63	37	0	57	0.89
MADCHOW-16	0	51	49	0	0.72
MADCHOW-17	0	100	0	0	1.00
MADCHOW-18	84	3	14	71	0.59
MADCHOW-19	12	88	0	12	0.99
MADCHOW-20	100	0	0	0	0.11
MADCHOW-21	24	63	13	20	1.00
MADCHOW-22	96	0	3	5	0.17
MADCHOW-23	96	2	2	0	0.05
MADCHOW-24	100	0	0	0	0.28
MADCHOW-25	99	1	0	0	0.24
MADCHOW-26	100	0	0	6	0.09
MADCHOW-27	96	0	4	92	0.25
MADCHOW-28	100	0	0	32	0.28
MADCHOW-29	77	22	1	77	0.11
MADCHOW-30	97	3	0	23	0.17
Understanding wells					
MADCHOWFP-01	95	0	4	14	0.18
MADCHOWFP-02	75	8	17	58	0.63
MADCHOWFP-03	100	0	0	0	0.04
MADCHOWFP-04	97	3	0	64	0.81
MADCHOWFP-05	47	53	0	0	0.88

[1] Percent agricultural plus percent natural plus percent urban land use add up to 100 percent. Orchard/vineyard land use is a subset of agricultural land use.

Table A2.   Well construction characteristics and position relative to the Corcoran Clay for grid and understanding wells sampled in April and May 2008, Madera-Chowchilla study unit, 2008, California GAMA Priority Basin Project.

[USGS-GAMA well identification number: MADCHOW, Madera-Chowchilla study unit grid well; MADCHOWFP, Madera-Chowchilla study unit understanding wells. Other abbreviations: na, not available; >, greater than]

USGS-GAMA well identification number	Construction information (feet below land surface datum, except where noted)				Position relative to Corcoran Clay
	Well depth	Top of perforations	Bottom of perforations	Length of perforated interval, in feet	
Grid wells					
MADCHOW-01	592	210	588	378	East
MADCHOW-02	600	240	600	360	East
MADCHOW-03	540	240	520	280	East
MADCHOW-04	480	180	470	290	East
MADCHOW-05	350	290	350	60	East
MADCHOW-06	310	48	310	262	East
MADCHOW-07	>300	na	na	na	East
MADCHOW-08	820	410	800	390	East
MADCHOW-09	234	234	234	0	Below
MADCHOW-10	830	506	830	324	Below
MADCHOW-11	780	385	770	385	East
MADCHOW-12	300	240	300	60	Below
MADCHOW-13	670	500	660	160	East
MADCHOW-14	388	388	388	0	East
MADCHOW-15	450	390	450	60	East
MADCHOW-16	740	330	740	410	East
MADCHOW-17	140	na	na	na	East
MADCHOW-18	330	210	280	70	East
MADCHOW-19	200	140	200	60	East
MADCHOW-20	352	200	340	140	Above/Across
MADCHOW-21	320	240	320	80	East
MADCHOW-22	325	325	325	0	Below
MADCHOW-23	655	400	655	255	Below
MADCHOW-24	294	244	294	50	Below
MADCHOW-25	>200	na	na	na	Unknown
MADCHOW-26	510	210	510	300	Above/Across
MADCHOW-27	480	240	480	240	East
MADCHOW-28	216	204	212	8	Above/Across
MADCHOW-29	340	160	324	164	Above/Across
MADCHOW-30	388	358	388	30	Below
Understanding wells					
MADCHOWFP-01	254	212	254	42	Above/across
MADCHOWFP-02	377	242	377	135	Below
MADCHOWFP-03	198	163	198	35	Above/across
MADCHOWFP-04	200	184	196	12	East
MADCHOWFP-05	340	240	340	100	East

Tritium enters the hydrologic cycle following oxidation to tritiated water. Natural background levels of tritium in precipitation are approximately 3 to 15 TU (Craig and Lal, 1961; Clark and Fritz, 1997). Above-ground nuclear explosions resulted in a large increase in tritium values in precipitation, beginning in about 1950 and peaking in 1963 at values over 1,000 TU in the northern hemisphere (Michel, 1989). Radioactive decay over a period of 60 years would decrease tritium values of 10 TU to 0.6 TU.

Previous investigations have used a range of tritium values from 0.2 to 1.0 TU as thresholds for indicating presence of water that has exchanged with the atmosphere since about 1950 (Michel, 1989; Plummer and others, 1993; Michel and Schroeder, 1994; Clark and Fritz, 1997; Manning and others, 2005; Landon and others, 2010; Kulongoski and others, 2010). For samples collected for the Madera-Chowchilla study unit in 2008, tritium values greater than a threshold of 0.2 TU were defined as indicating presence of groundwater recharged since about 1950. By using a tritium value of 0.2 TU for the threshold in this study, the age classification scheme allows for samples with a slightly larger fraction of pre-modern groundwater to be classified as modern than if a higher threshold were used. A higher threshold for tritium would result in fewer samples classified as modern than classified as pre-modern, when carbon-14 would suggest that the samples were primarily modern.

Carbon-14 (^{14}C) is a widely used chronometer based on the radiocarbon content of organic and inorganic carbon. Dissolved inorganic carbonate species typically are used for ^{14}C dating of groundwater. Carbon-14 is formed in the atmosphere by the interaction of cosmic-ray neutrons with nitrogen, and to a lesser degree, with oxygen and carbon. Carbon-14 is incorporated into carbon dioxide and mixed throughout the atmosphere. The carbon dioxide dissolves in precipitation which eventually recharges the aquifer. Carbon-14 activity in groundwater, expressed as percent modern carbon (pmc), reflects exposure to the atmospheric ^{14}C source and is governed by the decay constant of ^{14}C (with a half-life of 5,730 yrs). Carbon-14 can be used to estimate groundwater ages ranging from 1,000 to approximately 30,000 years before present because of its half-life. Calculated ^{14}C ages in this study are referred to as "uncorrected" because they have not been adjusted to consider exchanges with sedimentary sources of carbon (Fontes and Garnier, 1979). The ^{14}C age (residence time) is calculated on the basis of the decrease in ^{14}C activity because of radioactive decay since groundwater recharge, relative to an assumed initial ^{14}C concentration (Clark and Fritz, 1997). An average initial ^{14}C activity of 99 percent modern carbon (pmc) is assumed for this study, with estimated errors on calculated groundwater ages up to ± 20%. Groundwater with a ^{14}C activity of >88 pmc is reported as having an age of <1,000 years; no attempt is made to refine ^{14}C ages <1,000 years. Measured values of percent modern carbon can be > 100 pmc because the definition of the ^{14}C activity in "modern" carbon does not include the excess ^{14}C produced in the atmosphere by above-ground nuclear weapons testing. For the Madera-Chowchilla study unit, ^{14}C activity ≤80 pmc was defined as indicative of presence of groundwater recharged before about 1950. The threshold value of 80 pmc was selected because all groundwater samples with tritium ≤ 0.2 TU also had ^{14}C ≤80 pmc.

Helium (He) is a naturally occurring inert gas initially included during accretion of the planet, and later produced by radioactive decay of lithium, uranium, and thorium in the Earth. Helium (^{3}He plus ^{4}He) concentrations in groundwater often exceed the expected solubility equilibrium values as a result of air-bubble entrainment, or subsurface production of both isotopes, and their subsequent release into the groundwater (for example, Morrison and Pine, 1955; Andrews and Lee, 1979; Torgersen, 1980; Torgersen and Clarke, 1985). There are four primary sources of He in groundwater:

$$He_{total} = He_{equil} + He_{exair} + He_{trit} + He_{terr},$$

where

He_{total} is the total amount of helium in the groundwater sample;

He_{equil} is the helium derived from equilibration with the atmosphere at the time of recharge;

He_{exair} is the helium derived from dissolved air bubbles ("excess" air);

He_{trit} is the helium produced by radioactive decay of tritium in the sample; and

He_{terr} is the helium produced by radioactive decay of uranium and thorium in aquifer material or emanating from deeper in the Earth's crust or mantle.

He_{equil}, He_{exair}, and He_{terr} all consist of helium-3 (^{3}He) and helium-4 (^{4}He); however, He_{trit} consists only of ^{3}He.

He_{equil} is a function of temperature at the time of recharge. Recharge temperatures were calculated from dissolved neon, argon, krypton, and xenon using methods described in Aeschbach-Hertig and others (1999, 2000) to model the He_{exair} component. The best model for the He_{exair} component for each groundwater sample was selected by comparing the sums of the weighted squared standard deviations between the modeled and measured noble-gas concentrations ($\chi 2$). The model with the lowest $\chi 2$ value (least amount of deviation between the modeled and measured concentrations) was selected. The $\chi 2$ was compared to the value of a chi-squared distribution with the appropriate number of degrees of freedom for the model and a significance level (α) of 0.01 ($\chi2\alpha=0.01$). Recharge temperatures were only calculated for groundwater samples for which $\chi2$ was less than $\chi2\alpha=0.01$ (Aeschbach-Hertig and others, 2000).

The presence of large concentrations of He_{terr} may be indicative of long groundwater residence times. For the purpose of estimating groundwater residence times, the amount of He_{terr} is converted to the parameter $\%He_{terr-c}$, the percent of He_{terr} in He_{total} corrected for excess air:

$$\%He_{terr-c} = \frac{He_{terr}}{He_{total-}He_{exair}} \times 100.$$

(He_{trit} is neglected in calculation of $\%He_{terr-c}$ because it typically is very small.) For the Madera-Chowchilla study units, values of $\%He_{terr-c} > 10\%$ were defined as indicative of the presence of pre-modern groundwater. The threshold of 10% was selected because many of the samples with ^{14}C ≤ 80 pmc also had $\%He_{terr-c} > 10\%$.

The $^3He/^4He$ ratio of He_{terr} was determined by the linear regression of He_{terr}/He_{total} and δ^3He [($\delta^3He = R_{meas}/R_{atm} -1$) $\times$ 100 percent, where R_{meas}/R_{atm} is the measured $^3He/^4He$ ratio in the sample divided by the $^3He/^4He$ ratio in the atmosphere] for groups of related groundwater samples with tritium <1 TU.

Noble gas concentrations and measured $^3He/^4He$ ratios are reported in table A3, and tritium, ^{14}C ages, and $\%He_{terr-c}$ are reported in table A4. Because of uncertainties in age distributions, particularly the uncertainties caused by mixing of waters of different ages in wells with long screened or open intervals and high withdrawal rates, the uncorrected ^{14}C ages were not specifically used for quantifying the relation between age and water quality in this report. While more sophisticated lumped parameter models for analyzing age distributions that incorporate mixing are available (Cook and Böhlke, 2000), use of these alternative models to understand age mixtures was beyond the scope of this report.

Groundwater samples with tritium >0.2 TU, ^{14}C >80 pmc, and $\%He_{terr-c} <10\%$ were classified as "Modern"; samples with tritium >0.2 TU, and $^{14}C \leq 80$ pmc or $\%He_{terr-c}$ $\geq 10\%$ were classified as "Mixed"; and samples with tritium ≤ 0.2 TU, and $^{14}C \leq 80$ pmc or $\%He_{terr-c} > 10\%$ were classified as "Pre-modern." Groundwater age classifications are reported in table A4. Classification into modern, mixed, and pre-modern categories was sufficient to provide an appropriate and useful characterization for the purposes of examining groundwater quality.

Geochemical Conditions

Geochemical conditions investigated as potential explanatory factors in this report include oxidation-reduction characteristics, pH, and cation ratios. Oxidation-reduction (redox) conditions influence the mobility of many organic and inorganic constituents (McMahon and Chapelle, 2008). Redox conditions along groundwater flow paths commonly proceed along a well-documented sequence of Terminal Electron Acceptor Processes (TEAP), in which a single TEAP typically dominates at a particular time and aquifer location (Chapelle and others, 1995; Chapelle, 2001). The predominant TEAPs are oxygen-reducing (oxic), nitrate-reducing, manganese-reducing, iron-reducing, sulfate-reducing, and methanogenic. The presence of redox-sensitive chemical species suggesting more than one TEAP may indicate mixing of waters from different redox zones upgradient of the well, that the well is perforated or screened across more than one redox zone, or spatial heterogeneity in microbial activity in the aquifer. Redox conditions were represented by dissolved oxygen concentration and by classified redox condition (table A5). Dissolved oxygen (DO) concentrations were measured at 34 of the 35 USGS-GAMA wells. Classifications of redox condition were made using the framework of McMahon and Chapelle (2008). An automated workbook program was used to assign the redox classification to each sample (Jurgens and others, 2009b). A key component to the accurate classification of redox conditions using the McMahon and Chapelle framework is availability of DO data, which was lacking in one well sampled in the Madera-Chowchilla study unit. Because the iron, manganese, and sulfate values were low in that well, the DO was estimated to be equal to or greater than 1.0; therefore, all 35 grid and understanding wells had enough information to make a determination of redox condition (table A5). Higher concentrations of iron, manganese, and sulfate constituents (greater than the threshold value used by the McMahon and Chapelle redox framework) are generally indicative of reducing conditions.

The primary cations in groundwater typically are calcium, magnesium, sodium, and potassium. For this study, the cation composition of groundwater was represented by the ratio of calcium+magnesium to the sum of all four cations, in units of milliequivalents per liter. This parameter is called Fract-CaMg, and is equivalent to the distance from the sodium plus potassium apex on the cation triangle portion of a Piper plot (Piper, 1944).

Table A3.   Results for analyses of noble gases in samples collected for the Madera-Chowchilla study unit, California GAMA Priority Basin Project.

[The five-digit number in parentheses below the constituent name is the U.S. Geological Survey parameter code used to uniquely identify a specific constituent or property. **USGS-GAMA well identification number:** MADCHOW, Madera-Chowchilla study unit grid well; MADCHOWFP, Madera-Chowchilla study unit understanding well. **Other abbreviations:** cm^3 STP-g^{-1} H_2O, cubic centimeters at standard temperature and pressure per gram of water; na, not available]

GAMA identification number	Sample collection data	Dissolved gas analysis date	Helium-3/ Helium-4 (atom ratio) (61040) $\times 10^{-7}$	Helium-4 (cm^3 STP-g^{-1} H_2O) (85561) $\times 10^{-7}$	Neon (cm^3 STP-g^{-1} H_2O) (61046) $\times 10^{-7}$	Argon (cm^3 STP-g^{-1} H_2O) (85563) $\times 10^{-4}$	Krypton (cm^3 STP-g^{-1} H_2O) (85565) $\times 10^{-8}$	Xenon (cm^3 STP-g^{-1} H_2O) (85567) $\times 10^{-9}$
Grid wells								
MADCHOW-01	04-14-08	12-11-09	31.67	0.54	2.34	3.25	6.78	9.14
MADCHOW-02	04-15-08	01-19-09	11.17	0.64	2.50	3.19	6.66	9.11
MADCHOW-03	04-15-08	01-19-09	16.12	0.61	2.38	3.49	7.59	9.96
MADCHOW-04	04-16-08	01-20-09	15.29	0.55	1.93	3.17	7.28	10.03
MADCHOW-05	04-16-08	01-20-09	7.10	1.77	2.63	3.18	6.67	8.90
MADCHOW-06	04-17-08	01-20-09	4.72	5.83	4.89	4.39	8.08	9.42
MADCHOW-07	04-21-08	01-06-10	12.50	0.55	2.00	3.01	6.26	8.82
MADCHOW-08	04-22-08	01-20-09	6.48	8.66	2.00	2.98	6.59	9.12
MADCHOW-09	04-22-08	12-04-09	16.25	0.47	3.80	3.15	5.69	8.82
MADCHOW-10	04-24-08	01-20-09	7.91	1.63	2.11	3.29	7.23	9.97
MADCHOW-11	04-24-08	01-20-09	10.20	4.84	8.74	6.52	10.79	11.91
MADCHOW-12	04-28-08	na	na	na	na	na	na	na
MADCHOW-13	04-29-08	01-21-09	11.63	0.69	2.28	3.36	7.26	9.95
MADCHOW-14	04-30-08	01-21-09	12.48	0.65	2.32	3.44	7.38	10.09
MADCHOW-15	04-30-08	12-11-09	5.25	12.89	2.18	3.24	7.20	10.15
MADCHOW-16	05-01-08	01-21-09	8.06	1.64	1.94	3.01	6.56	9.23
MADCHOW-17	05-01-08	01-28-09	10.79	0.94	1.96	2.91	6.33	7.82
MADCHOW-18	05-06-08	na	na	na	na	na	na	na
MADCHOW-19	05-06-08	01-21-09	4.81	2.76	2.40	3.38	7.42	10.31
MADCHOW-20	05-07-08	01-06-10	13.49	0.64	2.50	3.42	6.95	9.68
MADCHOW-21	05-07-08	01-21-09	11.80	0.56	1.93	2.98	6.48	8.54
MADCHOW-22	05-08-08	01-21-09	7.43	2.12	2.84	3.40	7.31	10.13
MADCHOW-23	05-12-08	12-14-09	6.16	12.28	2.18	3.33	7.51	9.92
MADCHOW-24	05-13-08	01-21-09	9.82	1.50	2.15	3.20	7.10	9.53
MADCHOW-25	05-13-08	01-22-09	13.11	0.78	2.63	3.63	7.38	9.69
MADCHOW-26	05-14-08	01-22-09	9.02	1.75	3.14	3.64	7.48	9.95
MADCHOW-27	05-14-08	01-22-09	32.30	0.63	2.43	3.40	7.41	9.77
MADCHOW-28	05-19-08	01-06-09	19.26	0.66	2.88	3.70	7.55	9.52
MADCHOW-29	05-20-08	01-06-10	15.22	0.73	2.84	3.62	7.37	10.34
MADCHOW-30	05-21-08	01-06-10	7.04	6.22	1.90	2.86	6.22	8.75
Understanding wells								
MADCHOWFP-01	04-23-08	12-17-09	18.57	0.55	2.34	3.37	7.05	9.79
MADCHOWFP-02	05-05-08	01-21-09	11.41	0.79	2.11	3.29	7.51	10.27
MADCHOWFP-03	05-15-08	01-06-09	14.25	0.91	2.86	3.63	7.39	9.74
MADCHOWFP-04	05-21-08	01-07-09	7.44	4.93	2.20	3.22	7.06	10.07
MADCHOWFP-05	05-22-08	01-06-10	8.57	0.76	2.13	2.95	6.17	8.78

Table A4. Summary of groundwater age data and classification into modern, mixed, and pre-modern age categories for samples collected for the Madera-Chowchilla study unit, 2008, California GAMA Priority Basin Project.

[USGS-GAMA well identification number: MADCHOW, Madera-Chowchilla study unit grid well; MADCHOWFP, Madera-Chowchilla study unit understanding well. Other abbreviations: TU, tritium units; na, data not available; >, greater than; <, less than]

USGS-GAMA well identification number	Tritium activity (TU)	Carbon-14 (percent modern)	Uncorrected carbon-14 age (years)	Terrigenic helium (percent of total helium)	Age classification
Grid wells					
MADCHOW-01	6.2	121	<1,000	0	Modern
MADCHOW-02	0.18	79	1,790	0	Pre-modern
MADCHOW-03	1.0	81	1,610	2.5	Modern
MADCHOW-04	4.0	106	<1,000	16	Mixed
MADCHOW-05	0.34	77	2,020	71	Mixed
MADCHOW-06	1.7	104	<1,000	91	Mixed
MADCHOW-07	0.40	53	5,050	10	Mixed
MADCHOW-08	0.31	50	5,420	95	Mixed
MADCHOW-09	5.2	116	<1,000	0	Modern
MADCHOW-10	0.09	65	3,420	71	Pre-modern
MADCHOW-11	0.03	45	6,250	85	Pre-modern
MADCHOW-12	0.25	7	21,310	na	Mixed
MADCHOW-13	0.37	62	3,750	21	Mixed
MADCHOW-14	0.31	84	1,310	13	Mixed
MADCHOW-15	0.15	38	7,740	97	Pre-modern
MADCHOW-16	0.12	59	4,200	73	Pre-modern
MADCHOW-17	3.1	112	<1,000	51	Mixed
MADCHOW-18	7.1	115	<1,000	na	Modern
MADCHOW-19	0.84	72	2,560	83	Mixed
MADCHOW-20	0.40	109	<1,000	1.7	Modern
MADCHOW-21	0.03	69	2,860	15	Pre-modern
MADCHOW-22	0.15	17	14,330	76	Pre-modern
MADCHOW-23	-0.06	5	24,110	96	Pre-modern
MADCHOW-24	1.5	55	4,750	69	Mixed
MADCHOW-25	1.1	115	<1,000	19	Mixed
MADCHOW-26	0.25	102	<1,000	68	Mixed
MADCHOW-27	8.5	99	<1,000	3.7	Modern
MADCHOW-28	6.8	115	<1,000	0	Modern
MADCHOW-29	2.7	109	<1,000	0	Modern
MADCHOW-30	0.34	22	12,010	93	Mixed
Understanding wells					
MADCHOWFP-01	5.0	101	<1,000	0	Modern
MADCHOWFP-02	0.25	87	1,000	38	Mixed
MADCHOWFP-03	2.7	106	<1,000	27	Mixed
MADCHOWFP-04	2.4	98	<1,000	91	Mixed
MADCHOWFP-05	0.15	11	17,490	34	Pre-modern

Table A5. Geochemical indicators for grid and understanding wells sampled in April and May 2008, Madera-Chowchilla study unit, 2008, California GAMA Priority Basin Project.

[**USGS-GAMA well identification number:** MADCHOW, Madera-Chowchilla study unit grid well; MADCHOWFP, Madera-Chowchilla study unit understanding well; **Redox information:** redox, oxidation-reduction; oxic, dissolved oxygen >0.5; anoxic (Mn), dissolved oxygen <0.5 and manganese (IV) reducing. **Other abbreviations:** mg/L, milligrams per liter; na, not analyzed; Ca, calcium; Mg, magnesium; Fract-CaMg, calcium plus magnesium in milliequivalents divided by sum of calcium, magnesium, sodium, and potassium in milliequivalents]

USGS-GAMA well identification number	Dissolved oxygen concentration, (mg/L)	pH, standard units	Redox classification	Fract-CaMg
Grid wells				
MADCHOW-01	5.1	6.8	Oxic	0.73
MADCHOW-02	6.5	7.3	Oxic	0.55
MADCHOW-03	3.3	7.2	Oxic	0.55
MADCHOW-04	1.4	7.3	Oxic	0.66
MADCHOW-05	na	7.0	Oxic	0.58
MADCHOW-06	5.7	7.8	Oxic	0.49
MADCHOW-07	4.5	7.3	Oxic	0.66
MADCHOW-08	1.7	7.5	Oxic	0.61
MADCHOW-09	7.7	7.1	Oxic	0.79
MADCHOW-10	4.5	7.7	Oxic	0.49
MADCHOW-11	7.5	7.5	Oxic	0.59
MADCHOW-12	0.3	8.0	Anoxic (Mn)	0.29
MADCHOW-13	4.5	7.7	Oxic	0.28
MADCHOW-14	6.1	7.7	Oxic	0.58
MADCHOW-15	1.4	7.1	Oxic	0.43
MADCHOW-16	4.2	7.4	Oxic	0.53
MADCHOW-17	5.7	7.2	Oxic	0.78
MADCHOW-18	7.2	6.8	Oxic	0.66
MADCHOW-19	4	7.0	Oxic	0.64
MADCHOW-20	7.1	7.1	Oxic	0.72
MADCHOW-21	6.7	7.0	Oxic	0.60
MADCHOW-22	1.2	8.0	Oxic	0.54
MADCHOW-23	0.7	8.4	Oxic	0.16
MADCHOW-24	5.7	8.0	Oxic	0.55
MADCHOW-25	5.7	7.0	Oxic	0.78
MADCHOW-26	6	7.0	Oxic	0.72
MADCHOW-27	1.6	7.6	Oxic	0.56
MADCHOW-28	7.5	6.9	Oxic	0.78
MADCHOW-29	7.2	7.2	Oxic	0.69
MADCHOW-30	3	8.4	Oxic	0.33
Understanding wells				1.00
MADCHOWFP-01	8.2	7.4	Oxic	0.74
MADCHOWFP-02	6	7.5	Oxic	0.61
MADCHOWFP-03	7.8	6.8	Oxic	0.76
MADCHOWFP-04	3.3	7.4	Oxic	0.58
MADCHOWFP-05	1.3	7.6	Oxic	0.65

Appendix B. Comparison of CDPH and USGS-GAMA Data

Well Depths

Of the 35 the wells sampled by USGS-GAMA, 21 were wells listed in the CDPH database. Of the remaining 14 wells, 7 were irrigation wells, and 7 were domestic wells. The assessments presented in this report are intended to characterize groundwater quality in the primary aquifer system, which is defined by the depth intervals over which wells listed in the CDPH database are open or perforated; therefore, it was important to verify that the irrigation and domestic wells sampled were open or perforated in the same depth ranges as were wells in the CDPH database. Well construction information is not available in the CDPH database, so the verification was made using information for the sampled wells only.

The 21 CDPH wells sampled by USGS-GAMA were divided into two types for comparison with the irrigation and domestic wells. Eight were CDPH wells serving drinking water to more than 4,000 people. These wells included wells operated by the larger cities in the study unit (Chowchilla, Madera, and Fresno) and wells operated by institutions with large resident populations (prisons). Thirteen were CDPH wells serving drinking water to less than 500 people. These wells included wells operated by schools, parks, and small businesses.

The median depth of CDPH wells serving large populations (670 ft) was significantly greater than the median depths of CDPH wells serving small populations (325 ft; $p=0.002$), irrigation wells (346 ft; $p=0.008$), or domestic wells (317 ft; $p=0.014$) (fig. B1A). There were no significant differences in the median depths of CDPH wells serving small populations, irrigation wells, and domestic wells. Median depths to the top of perforations for the four types of wells were similar, with the only significant difference being between CDPH wells serving large populations (286 ft) and irrigation wells (200 ft) ($p=0.019$; fig. B1B). The median length of perforation interval of CDPH wells serving large populations (342 ft) was significantly greater than the median length of perforation interval of CDPH wells serving small populations (60 ft; $p=0.001$), irrigation wells (146 ft; $p=0.040$), or domestic wells (40 ft; $p=0.003$) (fig. B1C). There were no significant differences in the median lengths of perforation intervals of CDPH wells serving small populations, irrigation wells, and domestic wells. These comparisons indicate that these irrigation and domestic wells were perforated over the same depth intervals as the CDPH wells serving small populations, thus, these irrigation and domestic wells can be considered representative of the primary aquifer system.

Hydrochemical Facies

Major ion data for grid wells were compared with major ion data from all wells in the CDPH database for this study unit to evaluate whether the grid wells were representative of the range of groundwater types used for public supply. The datasets were compared using Piper diagrams (Piper, 1944; Hem, 1992). Piper diagrams show the relative contribution of major cations and anions (on a charge equivalent basis) as a percentage of the total ion content of the water (fig. B2). All recent (February 12, 2005–February 12, 2008) CDPH data having cation/anion data and an acceptable cation/anion balance were retrieved and plotted on these Piper diagrams for comparison with data from the USGS-GAMA samples. Cation/anion balance was calculated as the absolute value of the difference between the total cations and total anions divided by the average of the total cations and total anions, expressed as a percentage, and acceptable cation-anion balance was defined as $<10\%$.

Similar ranges of water types were evident from grid wells and recent CDPH data (fig. B2). The anion compositions of the majority of CDPH and grid wells were classified as bicarbonate-type waters (anion composition greater than 60% bicarbonate), and most of the remainder was classified as bicarbonate-chloride-type. A few samples had chloride-type anion compositions. The cation compositions of the majority of CDPH and grid wells were classified as mixed-type (magnesium-calcium-sodium/potassium) or calcium-sodium/potassium-type. A few samples had sodium/potassium-type cation compositions. The similarity of the range of relative abundance of major cations and anions in grid wells to that of the set of all CDPH wells indicates that the grid wells represent the diversity of water types present within the Madera-Chowchilla study unit.

The two differences between water chemistry data for grid and understanding wells and recent 3-year data for CDPH wells are (1) a higher proportion of the grid wells are at either end of the cation trend (at lowest and highest proportion sodium/potassium), and (2) on the anion triangle, the grid wells have a higher proportion that have slightly elevated sulfate proportion. Some of the grid wells that cause these differences are located in areas where there are few CDPH wells, generally in the western portion of the study unit.

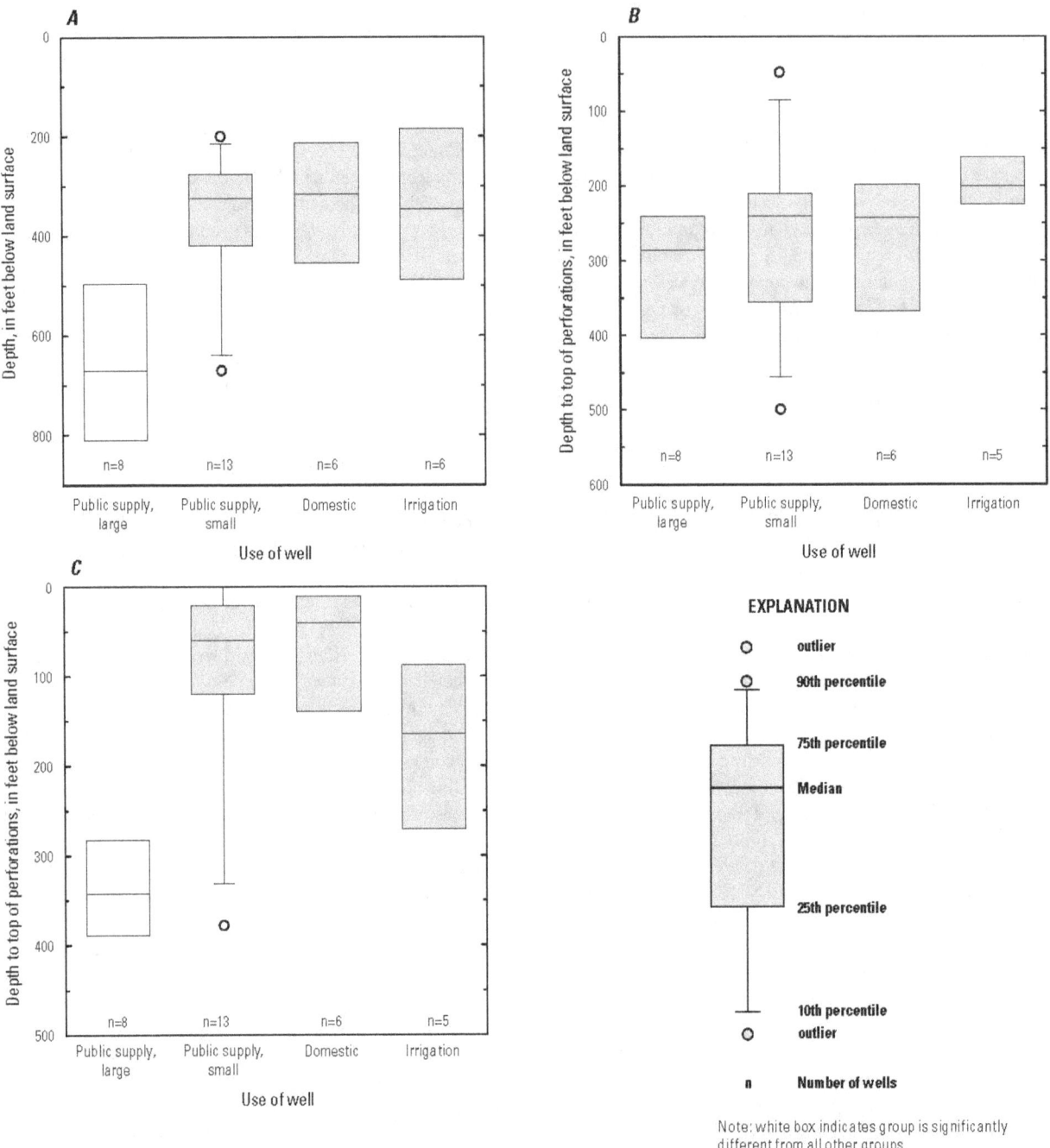

Figure B1.   Boxplots showing (*A*) well depth, (*B*) depth to top of perforations, and (*C*) perforation length for sampled grid and understanding wells, grouped by well type, Madera-Chowchilla study unit, California GAMA Priority Basin Project.

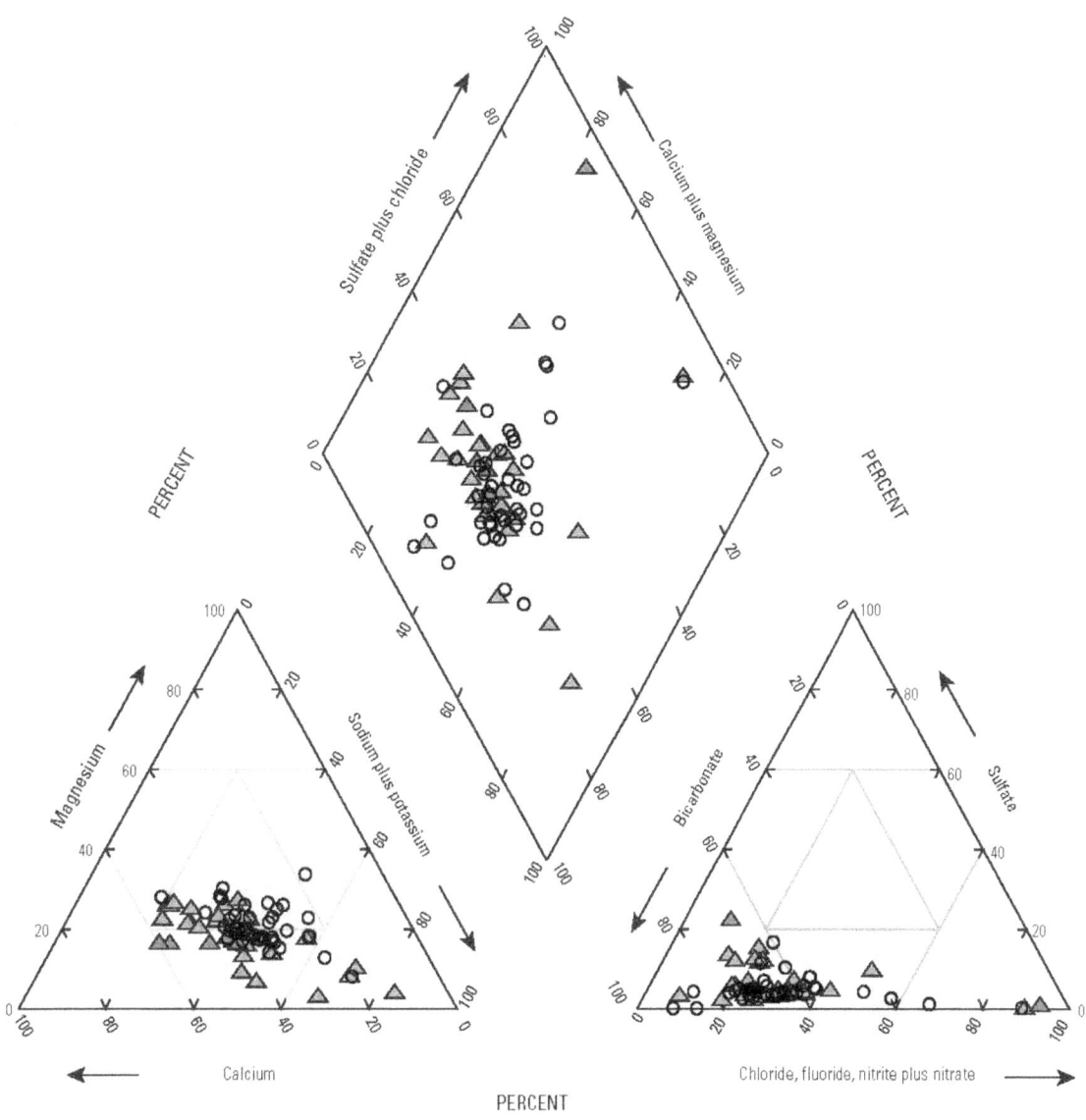

EXPLANATION

○ **CDPH well** (most recent analysis with charge imbalance less than 10 percent)

△ **grid well**

△ **understanding well**

Figure B2.   Piper diagram for grid and understanding wells and for wells in the California Department of Public Health database with data during February 12, 2005–February 12, 2008, Madera-Chowchilla study unit, California GAMA Priority Basin Project.

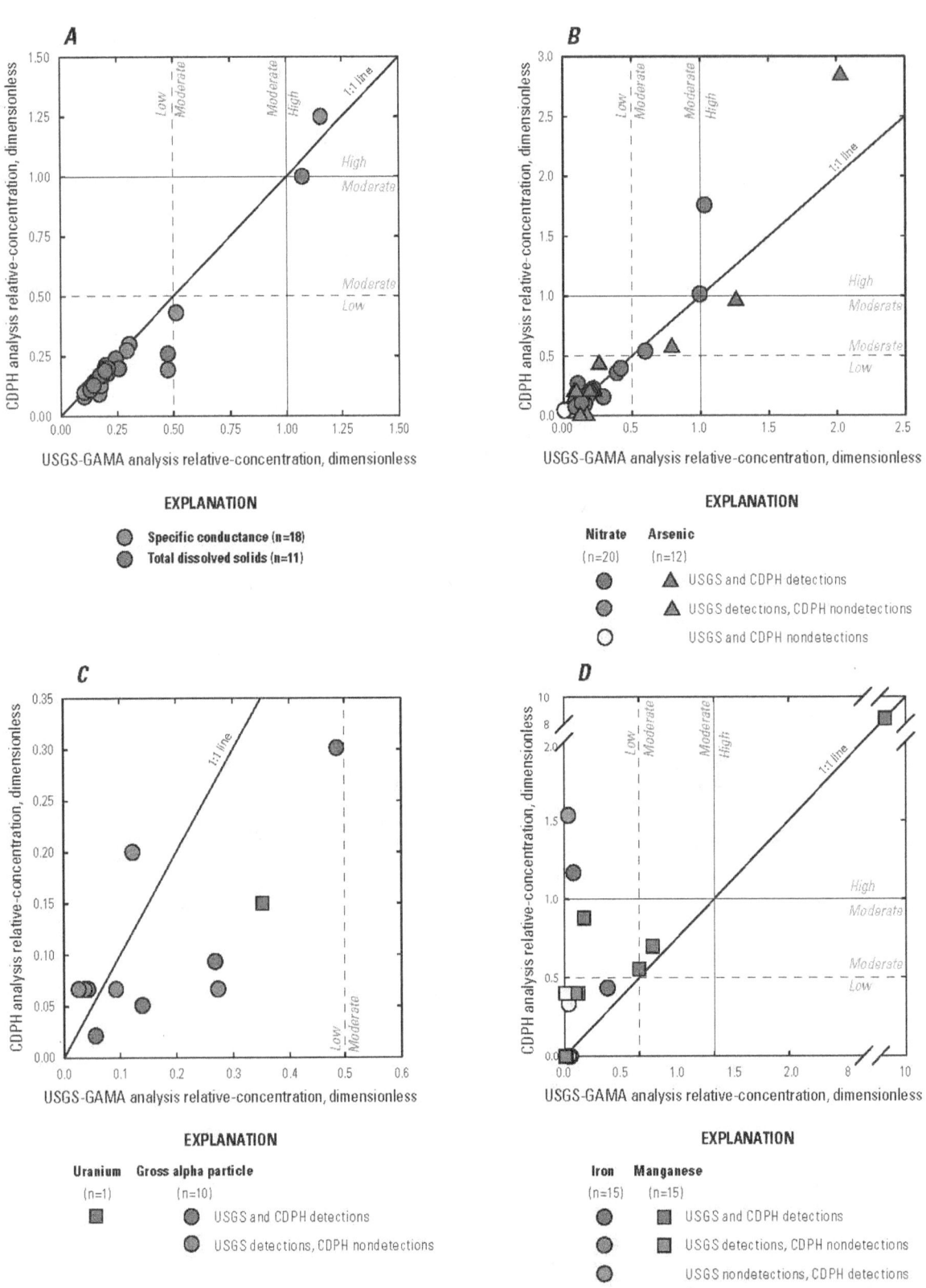

Figure B3.   Graphs showing relative-concentration values for (A) total dissolved solids and specific conductance, (B) arsenic and nitrate, (C) gross alpha particle and uranium activity, and (D) iron and manganese in groundwater from wells sampled by USGS-GAMA for the Madera-Chowchilla study unit that have data reported in the CDPH database between February 12, 2005, and February 12, 2008, California GAMA Priority Basin Project.

Water-Quality Constituents

Data acquired by USGS-GAMA and data from the CDPH database were compared to assess the validity of merging data from these two sources to create the dataset of inorganic and radioactive constituents used in the spatially weighted calculations of aquifer-scale proportions. Comparisons were made using data for the 20 grid and understanding wells that had data in the CDPH database between February 12, 2005, and February 12, 2008.

Eighteen wells had USGS-GAMA and CDPH data for specific conductance, and 11 wells had data for TDS from both sources (fig. B3A). USGS-GAMA specific conductance and TDS values were significantly higher than the CDPH values (Wilcoxon ranked-pair test; p = 0.012 and p = 0.011). The RC categories (high, moderate, or low) assigned to the wells were the same, however, whether the USGS-GAMA or CDPH values were used. Thus, although the datasets were significantly different, this difference was not large enough to affect the outcome of the *status assessment*.

Twenty wells had USGS-GAMA and CDPH data for nitrate, and 12 wells had data for arsenic from both sources (fig. B3B). There were no significant differences between USGS-GAMA and CDPH values for nitrate or arsenic (Wilcoxon ranked-pair test; p = 0.95 for nitrate; p = 0.9 for the 7 wells with arsenic detections in both USGS-GAMA and CDPH). The RC categories (high, moderate, or low) assigned to the wells were the same whether the USGS-GAMA or CDPH values were used.

Ten wells had USGS-GAMA and CDPH data for gross alpha particle activity, and 1 well had data for uranium activity from both sources (fig. B3C). For the five data pairs with detections in both datasets, USGS-GAMA values were higher than CDPH values, although the difference was not statistically significant because of the small number of pairs (Wilcoxon ranked-pair test; p = 0.059). The RC categories (high, moderate, or low) assigned to the wells were the same whether the USGS-GAMA or CDPH values were used.

Fifteen wells had USGS-GAMA and CDPH data for iron and manganese (fig. B3D). Three wells had detections reported in the CDPH database, and the USGS-GAMA results for those wells were either detections below the CDPH reporting limit of 100 µg/L or nondetections (laboratory reporting limit of 8 µg/L). Although the difference between the two datasets was not statistically significant due to the smallness of the datasets, the difference did have significant effects on the *status assessment* results. Iron was the only constituent for which the RC categories assigned to the wells by using USGS-GAMA and CDPH data were different: two wells that have low RCs calculated by using USGS-GAMA data have high RCs calculated by using CDPH data. These two wells, MADCHOW-05 and -09, have low RCs for manganese in both datasets. Thirteen CDPH wells had high RCs of iron reported between February 12, 2005, and February 12, 2008, and 9 of those 13 had low RCs of manganese. Given that iron reduction typically occurs at lower oxidation-reduction potentials than manganese reduction (Appelo and Postma, 2005; McMahon and Chapelle, 2008), one would not expect to detect elevated iron concentrations in groundwater that does not have elevated manganese concentrations. Between May 2004 and October 2010, USGS-GAMA collected samples from 87 wells statewide that had high RCs of iron; of these 87, only 3 had low RCs of manganese. The presence of high RCs of iron with low RCs of manganese in CDPH data but not in USGS data may be caused by differences in sample collection methods. USGS-GAMA trace element concentrations were measured in filtered groundwater samples (Shelton and others, 2009). CDPH trace element concentrations may have been measured on filtered or unfiltered groundwater samples. Suspended particles in unfiltered samples may contain iron if they are derived from metal parts of wells. For this reason, the high RCs of iron with low RCs of manganese are not considered representative of the water quality in the aquifer.